This book presents a powerful hybrid intelligent system based on fuzzy logic, neural networks, genetic algorithms, and related intelligent techniques. The new compensatory genetic fuzzy neural networks have been widely used in fuzzy control, nonlinear system modeling, compression of a fuzzy rule base, expansion of a sparse fuzzy rule base, fuzzy knowledge discovery, time series prediction, fuzzy games, and pattern recognition. The proposed soft computing system is effective in performing both linguistic-word-level fuzzy reasoning and numerical-data-level information processing. The book also presents various novel soft computing techniques.

Yanqing Zhang received his BSc and MSc in computer science and engineering from Tianjin University, China, in 1983 and 1986, respectively. He was conferred his PhD in computer science and engineering from the University of South Florida in 1997. From 1986 to 1993, he was a lecturer in the Department of Computer Science and Engineering at Tianjin University. Since 1997, he has been an assistant professor in the School of Computer and Applied Sciences at Georgia Southwestern State University. He is also a member of the IEEE, ACM and Upsilon Pi Epsilon.

Abraham Kandel received his BSc from the Technion-Israel Institute of Technology and MSc from the University of California, both in electrical engineering, and PhD in electrical engineering and computer science from the University of New Mexico. A professor and the Endowed Eminent Scholar in Computer Science and Engineering, he is Chairman of the Department of Computer Science and Engineering at the University of South Florida. He has published over 350 research papers and is a Fellow of the IEEE, ACM, AAAS and the New York Academy of Sciences.

SERIES IN MACHINE PERCEPTION AND ARTIFICIAL INTELLIGENCE*

Editors: **H. Bunke** (Univ. Bern, Switzerland)
P. S. P. Wang (Northeastern Univ., USA)

Vol. 17: Applications of AI, Machine Vision and Robotics
(Eds. *K. L. Boyer, L. Stark and H. Bunke*)

Vol. 18: VLSI and Parallel Computing for Pattern Recognition and AI
(Ed. *N. Ranganathan*)

Vol. 19: Parallel Image Analysis: Theory and Applications
(Eds. *L. S. Davis, K. Inoue, M. Nivat, A. Rosenfeld and P. S. P. Wang*)

Vol. 20: Picture Interpretation: A Symbolic Approach
(Eds. *S. Dance, T. Caelli and Z.-Q. Lin*)

Vol. 21: Modelling and Planning for Sensor Based Intelligent Robot Systems
(Eds. *H. Bunke, T. Kanade and H. Noltemeier*)

Vol. 22: Machine Vision for Advanced Production
(Eds. *M. Pietikainen and L. F. Pau*)

Vol. 23: Machine Learning and Perception
(Eds. *G. Tasini, F. Esposito, V. Roberto and P. Zingaretti*)

Vol. 24: Spatial Computing: Issues in Vision, Multimedia and Visualization Technologies
(Eds. *T. Caelli, Peng Lam and H. Bunke*)

Vol. 25: Studies in Pattern Recognition Memorial Volume in Honor of K S Fu
(Eds. *H. Freeman*)

Vol. 26: Neural Network Training Using Genetic Algorithms
(Eds. *L. C. Jain, R. P. Johnson and A. F. J. van Rooij*)

Vol. 27: Intelligent Robots — Sensing, Modeling & Planning
(Eds. *B. Bolles, H. Bunke and H. Noltemeier*)

Vol. 28: Automatic Bankcheck Processing
(Eds. *S. Impedovo, P. S. P. Wang and H. Bunke*)

Vol. 29: Document Analysis II
(Eds. *J. J. Hull and S. Taylor*)

Vol. 30: Compensatory Genetic Fuzzy Neural Networks and Their Applications
(*Y.-Q. Zhang and A. Kandel*)

Forthcoming

Introduction to Pattern Recognition — Statistical, Structural, Neural and Fuzzy Logic Approaches
(*A. Kandel and M. Friedman*)

*For the complete list of titles in this series, please write to the Publisher.

Series in Machine Perception and Artificial Intelligence – Vol. 30

COMPENSATORY GENETIC FUZZY NEURAL NETWORKS AND THEIR APPLICATIONS

Yanqing Zhang
University of South Florida
and
Georgia Southwestern State University

Abraham Kandel
University of South Florida
and
Tel-Aviv University

World Scientific
Singapore • New Jersey • London • Hong Kong

Published by

World Scientific Publishing Co. Pte. Ltd.
P O Box 128, Farrer Road, Singapore 912805
USA office: Suite 1B, 1060 Main Street, River Edge, NJ 07661
UK office: 57 Shelton Street, Covent Garden, London WC2H 9HE

Library of Congress Cataloging-in-Publication Data
Zhang, Yan-Qing.
    Compensatory genetic fuzzy neural networks and their applications
/ Yan-Qing Zhang, Abraham Kandel.
      p. cm. -- (Series in machine perception and artificial intelligence; vol. 30)
    Includes bibliographical references and index.
    ISBN 9810233493
    1. Neural networks (Computer science) 2. Fuzzy systems.
I. Kandel, Abraham. II. Title. III. Series
QA76.87.Z475   1997
006.3--dc21   97-44774
                                                                                                             CIP

British Library Cataloguing-in-Publication Data
A catalogue record for this book is available from the British Library.

Copyright © 1998 by World Scientific Publishing Co. Pte. Ltd.

All rights reserved. This book, or parts thereof, may not be reproduced in any form or by any means, electronic or mechanical, including photocopying, recording or any information storage and retrieval system now known or to be invented, without written permission from the Publisher.

For photocopying of material in this volume, please pay a copying fee through the Copyright Clearance Center, Inc., 222 Rosewood Drive, Danvers, MA 01923, USA. In this case permission to photocopy is not required from the publisher.

This book is printed on acid-free paper.

Printed in Singapore by Uto-Print

To Mingyi, Zhifeng, Jingju, Yonglu,
Sharon, Gill and Adi.

Preface

In recent years, fuzzy systems, neural networks and genetic algorithms have attracted more and more interest of scientists, engineers, researchers and students in various multi-disciplinary fields of science and technology. The main reason for the growing interest is that these different techniques can be merged into a hybrid intelligent system so called *a soft computing system* capable of not only discovering inexact knowledge in *crisp* or fuzzy data but also generating *linguistic*-level approximation via approximate reasoning. Although various soft computing systems have been developed for a number of applications, "how to design a soft computing system by effectively incorporating fuzzy systems, neural networks, genetic algorithms and other relevant techniques" still is a challenging issue. For this important subject, this monograph focuses on representing various novel hybrid systems integrating fuzzy logic, neural networks, genetic algorithms, compensatory operations and heuristic methods by improving conventional methods and discovering new principles.

This book is organized in three parts: the theoretical part consists of Chapters 1-4, the application part is composed of Chapters 5-11, and the conclusion part is Chapter 12. In the theoretical part, Chapter 1 generally reviews main issues related to fuzzy sets theory, artificial neural networks, genetic algorithms and soft-computing systems. Chapter 2 presents various kinds of compensatory fuzzy operations according to Yin-Yang compensatory principles, and extensively investigates the compensation between fuzzy DNF (Disjunctive Normal Form) and fuzzy CNF (Conjunctive Normal Form) to understand the natural divergence of a fuzzy logical statement. Chapter 3 demonstrates that (1) data granularity of conventionally used fuzzy sets is not appropriate to contain more heuristic information for effective fuzzy reasoning and (2) commonly used fuzzy reasoning methods may result in unreasonable behaviors under some circumstance. In order to discuss these issues, we have developed a compensatory fuzzy reasoning methodology using new primary fuzzy sets with appropriate data granularity according to the philosophical principles of Taichi and properties of increasing and decreasing functions. Chapter 4 is devoted to the design of a Fuzzy Neural Network with Knowledge Discovery (FNNKD) in order to perform the compensatory fuzzy reasoning. The FNNKD is more powerful and more efficient than either Takagi-Sugeno's fuzzy system or Jang's ANFIS because (1) all parameters in the FNNKD have physical meanings and therefore they can heuristically be initialized to speed up the learning based on training data and (2) the FNNKD is able to learn commonly used fuzzy IF-THEN rules from given data. Finally, a compensatory genetic fuzzy neural network using dedicated FNNKDs as basic building blocks is developed to process not only crisp input/output values but also fuzzy input/output sets. Compared with Wang's method, Jang's method and Sugeno-Kang's method, the compensatory genetic fuzzy neural network has impressive abilities of knowledge discovery and knowledge application. In the application part (Chapters 5-11), extensive discussion of illustrations regarding defuzzification techniques, a nonlinear function approximation, a cart-pole balancing system, compression of a fuzzy rule base, expansion of a sparse fuzzy rule base, highly nonlinear system modeling, a chaotic time series prediction, a gas furnace model identification, fuzzy games, fuzzy moves in Prisoner's Dilemma, pattern recognition and constructive methods to building fuzzy systems has strongly indicated that the compensatory genetic fuzzy neural network is an efficient and robust soft-computing system with abilities of discovering fuzzy knowledge from numerical data and applying trained fuzzy rules to perform complex nonlinear behaviors. In Chapter 12, major conclusions and future works are given.

We would like to acknowledge our indebtedness to all scientists and researchers who have contributed to the fast growing area of soft-computing incorporating fuzzy logic, neural networks, genetic algorithms and other conventional or novel techniques. We would like to thank Dr. Hall, Dr. Turksen, Dr. Friedman, Dr. Bunke, Dr. Ma, Dr. Goldgof, Dr. Khator, Dr. You, Dr. Anderson, Dr. Pedrycz, Dr. Jang and Dr. Cai and for their excellent comments. We would like to thank Adam Schenker for his proofreading an early version of this book.

Yan-Qing Zhang
Abraham Kandel

Tampa, Florida
September 1997

Contents

Preface vii

1 Introduction 1
   1.1 Fuzzy Sets and Data Granularity 1
   1.2 Neural Networks and Knowledge Discovery 3
   1.3 Genetic Algorithms and Adaptive Optimization 4
   1.4 Soft Computing Systems and Computational Intelligence 5
   1.5 Main Issues 7

2 Fuzzy Compensation Principles 9
   2.1 Fuzzy Yin-Yang Compensation 9
   2.2 Compensation of Fuzzy CNF and Fuzzy DNF 11
      2.2.1 Boolean Truth Table and Karnaugh Map 12
      2.2.2 Kaufmann's Fuzzy Truth Table and Fuzzy Map 13
      2.2.3 Universal Fuzzy Truth Table and AND_n^m Map 15
   2.3 2-variable-2-dimensional CNFs and DNFs 19
   2.4 2-variable-m-dimensional CNFs and DNFs for $m = 3, 4$ 21
   2.5 Compensation of Universal Fuzzy CNF and Fuzzy DNF 23
      2.5.1 Boolean Logic 23
      2.5.2 General Fuzzy Logic 26
      2.5.3 m-dimensional Fuzzy CNFs and DNFs of $a \wedge \bar{a}$ and $a \vee \bar{a}$ 29
      2.5.4 m-dimensional t-norm-t-conorm CNFs and DNFs 31
      2.5.5 Relations in Fuzzy and t-norm-t-conorm CNFs and DNFs 35

	2.6	Summary	40
3	**Normal Fuzzy Reasoning Methodology**		**41**
	3.1	Primary Fuzzy Subsets	41
	3.2	The Variable-Input-Constant-Output (VICO) Problem	42
	3.3	Normal Fuzzy Reasoning (NFR)	44
	3.4	Normal Fuzzy Controllers	49
4	**Compensatory Genetic Fuzzy Neural Networks**		**57**
	4.1	Introduction	57
	4.2	Fuzzy Neural Networks with Knowledge Discovery	58
	4.3	Heuristic Genetic Learning Algorithm for a FNNKD	61
	4.4	Feature Expressions of Trapezoidal-type Fuzzy Sets	67
	4.5	Crisp-Fuzzy Neural Networks (CFNN)	68
5	**Fuzzy Knowledge Rediscovery in Fuzzy Rule Bases**		**71**
	5.1	Applicability of Various Defuzzification Techniques	71
	5.2	Nonlinear Function Approximation	77
6	**Fuzzy Cart-pole Balancing Control Systems**		**81**
	6.1	Cart-pole Balancing Fuzzy Control Systems	81
	6.2	A Cart-pole Balancing System with Crisp Inputs and Outputs	86
	6.3	A Cart-pole Balancing System with Fuzzy Inputs and Outputs	90
7	**Fuzzy Knowledge Compression and Expansion**		**95**
	7.1	Compression of Fuzzy Rule Bases	95
	7.2	Expansion of Fuzzy Rule Bases	98
8	**Highly Nonlinear System Modeling and Prediction**		**104**
	8.1	Nonlinear Function Prediction	104
	8.2	Chaotic Time Series Prediction	106
		8.2.1 Wang's Fuzzy System and a FNNKD	108
		8.2.2 Effectiveness of the HGLA	109
		8.2.3 Analysis of Compensatory Degrees γ^k	110
		8.2.4 Performance of Various Approaches	111
	8.3	Box and Jenkins's Gas Furnace Model Identification	112

9 Fuzzy Moves in Fuzzy Games — 115
- 9.1 Introduction — 115
- 9.2 Fuzzy Moves — 116
- 9.3 Normal Fuzzy Reasoning for Fuzzy Moves — 118
- 9.4 Applicability of Various Methods — 119
  - 9.4.1 Prisoners' Dilemma — 119
  - 9.4.2 Applicability of Fuzzy Reasoning Methods — 123
- 9.5 Efficient Precise Decision Systems for Fuzzy Moves — 123
- 9.6 Typical Examples — 126
- 9.7 Fuzzy Moves in Prisoner's Dilemma — 127
  - 9.7.1 Global Games and Global PDs — 129
  - 9.7.2 Theory of Fuzzy Moves — 136
  - 9.7.3 Fuzzy Moves in Global PDs — 140
  - 9.7.4 Conclusions — 143
- 9.8 Summary — 144

10 Genetic Neuro-fuzzy Pattern Recognition — 145
- 10.1 Structure of a Genetic Fuzzy Neural Network — 145
- 10.2 Genetic-Algorithms-Based Self-Organizing Learning Algorithm — 147
- 10.3 Simulations — 149
- 10.4 Conclusions — 151

11 Constructive Approach to Modeling Fuzzy Systems — 152
- 11.1 Introduction — 152
- 11.2 A Normal-Fuzzy-Reasoning-Based Fuzzy System — 153
- 11.3 Various Single-Input-Single-Output (SISO) fuzzy systems — 154
- 11.4 Universal approximation — 157
- 11.5 A piecewise nonlinear constructive algorithm — 159
- 11.6 Simulations — 163
  - 11.6.1 A nonlinear function approximation — 163
  - 11.6.2 Box and Jenkins's gas furnace model identification — 165
  - 11.6.3 A chaotic system identification — 166
- 11.7 Conclusions — 168

12 Conclusions	**169**
12.1 Main Conclusions	169
12.2 Future Research and Development	171
Bibliography	**173**
Index	**183**

COMPENSATORY GENETIC FUZZY NEURAL NETWORKS AND THEIR APPLICATIONS

Chapter 1

Introduction

Hybrid intelligent systems are playing a more and more important role in a variety of applications of science and technology because they are built by merging different kinds of advanced techniques. In recent years, a powerful hybrid intelligent system called a soft computing system has intensively been investigated by researchers in both academia and industry, and has been widely used in various fields of complex systems. Generally, soft computing is a hybrid approach to solving complex problems with uncertainty and imprecision based on various computational intelligence techniques such as fuzzy logic, neural networks, genetic algorithms, probabilistic reasoning and other relevant techniques. Practically speaking, we need to investigate various relevant intelligent techniques separately, then solve existing problems behind them, and finally use improved intelligent techniques to design a powerful soft computing system. Therefore, several important techniques are analyzed in this chapter.

1.1 Fuzzy Sets and Data Granularity

Since Zadeh first defined the concept of fuzzy set in 1965 [111], fuzzy sets representing linguistic structures have been used as basic computational elements in fuzzy reasoning and in the processing of fuzzy information [20,24,26,42-44,58,66,67, 71-73,91-95,103-105,108-113,118]. Recently, Zadeh has investigated Computing with Words (CW) as described in several papers starting with [112] in which the concepts of a linguistic variable and granulation have been introduced. In CW, a granule is a fuzzy set of points having the form of a clump of elements drawn together by similarity [113]. In the sense of data granularity, data granularity of conventional fuzzy rule bases with fuzzy sets is much lower than that of databases with raw data. For example, a fuzzy rule base has 10 fuzzy numbers (i.e., big granules denoted by $\tilde{1}, \tilde{2}, ..., \tilde{10}$) defined in $[1,10]$, whereas a database has infinite real numbers (i.e., tiny granules) in $[1,10]$. In the sense of knowledge discovery, the degree of intelligence of a system using knowledge bases with lower data granularity is higher than that of a system using databases with higher data granularity since useful knowledge has been discovered from raw data. For the above example, we define a nonlinear function $f(x) = x^2$ in $[1,10]$. Because there exist infinite real numbers x in $[1,10]$, we have *infinite precise* rules in a database such that

IF X is x Then Y is x^2.

But we have only *10 fuzzy* rules in a fuzzy rule base such that

IF X is $\tilde{1}$ Then Y is $\tilde{1}$,

IF X is $\tilde{2}$ Then Y is $\tilde{4}$,

...

IF X is $\tilde{10}$ Then Y is $\tilde{100}$.

Obviously, it is much easier for a child to learn the 10 fuzzy rules than to remember a large number of precise rules. Actually, the general mapping relations (i.e., raw knowledge) imply the 10 finite ones (i.e., higher knowledge), as well as only one mapping relation $f(x) = x^2$ (i.e., perfect knowledge).

Naturally, we may have to deal with some relevant problems such as "How do we define condensed fuzzy granules?", "What is the optimal number of fuzzy granules?" and "How do we discover effective fuzzy rules based on given raw data?". For these problems, various techniques of knowledge acquisition and knowledge discovery have been used to extract knowledge with higher data granularity from raw data with the lowest data granularity in order to make a system more intelligent and more efficient [1,2,15-17,20,23,28,42-44,48,56,62,71-73,76-79,103,115]. However, the most important problem, which is "How to select optimal fuzzy granules with an appropriate degree of data granularity to make a system more intelligent and more efficient?", is still unsolved. In other words, commonly used fuzzy sets such as *young, not very young* do not have an appropriate data granularity. Since conventionally used fuzzy sets are the smallest blocks in a fuzzy rule base, conventional fuzzy techniques are at the level of middle fuzzy data granularity. Unfortunately, such middle fuzzy data granularity is not optimal because the conventional method may result in unreasonable behaviors in real applications. For example, in the fuzzy control problem (pp.161-166 in [61]), 5 fuzzy rules are used to control the motor speed of an air conditioner such that

Rule 1: If Temperature is cold, Motor Speed stops;

Rule 2: If Temperature is cool, Motor Speed slows;

Rule 3: If Temperature is just right, Motor Speed is medium;

Rule 4: If Temperature is warm, Motor Speed is fast;

Rule 5: If Temperature is hot, Motor Speed blasts.

The conventional fuzzy method results in the same air conditioner speed of 30 rpm with the increasing of temperature from $50°F$ to $60°F$, and so does the probabilistic approach [49,61]. Intuitively, the air conditioner speed should increase when temperature increases according to the five fuzzy rules. Through analyzing such a non-intuitive variable-input-constant-output phenomenon in Chapter 3, we have found the main reason for this phenomenon: the conventional fuzzy sets theory takes all commonly used fuzzy sets as fundamental elements for fuzzy reasoning without splitting some complex fuzzy sets into fuzzy subsets with heuristic information contained in a given fuzzy rule base. The data granularity of conventionally used fuzzy sets is too low to contain more heuristic information and more potential knowledge.

1.2 Neural Networks and Knowledge Discovery

A neural network has a massively parallel and distributed structure which is composed of many simple processing elements (i.e., artificial neurons) with nonlinear activation functions. The neurons in a neural network can communicate with each other through links between the neurons. Actually, an artificial neural network represents knowledge by distributed crisp weights. Unfortunately, a main disadvantage of a conventional crisp neural network is that crisp weights in the conventional crisp neural network are not directly related to parameters of fuzzy membership functions for linguistic terms in fuzzy IF-THEN rules, thus it is not convenient for us to use the crisp neural network to learn fuzzy IF-THEN rules from data.

In order to conveniently process fuzzy information and learn fuzzy knowledge, various fuzzy neural networks have been developed [15-21,28,31,334,35,39-44,46, 52,54,56,57,62,65,68,69,71-73,76,78,82]. A fuzzy neural network consists of many simple fuzzy neurons which perform various kinds of fuzzy operations. In the narrow sense, a fuzzy neural network is a fuzzy-operation-oriented neural network by directly fuzzifying inputs, weights, or outputs and using fuzzy operations such as fuzzy logical operations and fuzzy number operations [16,40]. The fuzzy-operation-oriented fuzzy neural network still can not conveniently extract fuzzy IF-THEN rules because fuzzified weights are not related to parameters of fuzzy membership functions for linguistic terms in fuzzy IF-THEN rules. Generally, the key problem is that crisp weights in a conventional crisp neural network and fuzzy weights in a fuzzy-operation-oriented neural network have no known explicit physical meaning related to fuzzy IF-THEN rules.

In order to easily learn fuzzy IF-THEN rules from data, fuzzy-reasoning-oriented neural networks with both adaptive learning and fuzzy reasoning abilities have been developed in recent years. For example, Jang's ANFIS [42-44] and Lin's Neural Fuzzy Control Systems (NFCS) [71,72] are such fuzzy-reasoning-oriented neural networks. Importantly, a fuzzy-reasoning-oriented neural network is capable of extracting high-level fuzzy knowledge such as fuzzy IF-THEN rules from data. .However, many existing fuzzy-reasoning-oriented fuzzy neural networks such as Jang's ANFIS and Lin's NFCS are Crisp-Input-Crisp-Output (CICO) models, so they can not directly be used in the other models which are the Crisp-Input-Fuzzy-Output (CIFO) model, the Fuzzy-Input-Crisp-Output (FICO) model and the Fuzzy-Input-Fuzzy-Output (FIFO) model. Thus, existing fuzzy-reasoning-oriented fuzzy neural networks are not suitable to some applications such as compressing a big fuzzy rule base or expanding a sparse fuzzy rule base. An important problem is how to design a universal fuzzy neural network incorporating CICO, CIFO, FICO and FIFO models.

1.3 Genetic Algorithms and Adaptive Optimization

In recent years, genetic algorithms have been widely used in many applications [1,2,9,10, 17,23,25,27,28,31-33,36,37,41,62,90,115]. In essence, genetic algorithms are adaptive procedures of optimization and search, based on the mechanics of natural selection and natural genetics [33,37]. Because genetic algorithms are robust and simple search techniques in general, they play an important role in optimizing a soft computing system with a large number of parameters. Many genetic-algorithms-based soft computing systems have been developed for different applications. In most cases, genetic algorithms can only find near-optimal solutions to an optimization problem because of randomness of genetic algorithms. These solutions may be satisfactory for some applications, but not acceptable for others. In this sense, a search method using only genetic algorithms is not powerful enough in some complex applications. Intuitively, a hybrid search algorithm based on genetic algorithms and some other search techniques can get a much better solution than genetic algorithms alone. Therefore, we have to design a hybrid learning algorithm based on genetic algorithms and some other heuristic algorithms for a fuzzy neural network with a large number of parameters.

1.4 Soft Computing Systems and Computational Intelligence

In the 1970's, Lee, Lee and Kandel [46,68,69] were the first to introduce the theory of fuzzy sets to the conventional McCulloch-Pitts model and study fuzzy neural networks based on the principle of neural networks and the mechanism of fuzzy automata. However, the development of research in neuro-fuzzy systems was very slow since (1) there were few researchers who did work on either neural networks or fuzzy logic systems and (2) the researchers didn't design powerful learning algorithms for neural networks and didn't have real applications for fuzzy logic systems. After a relatively slow period with neural networks and fuzzy logic in the 1970's, neural networks and fuzzy logic systems attracted a resurgence of attention from many researchers in a variety of scientific and engineering areas in the 1980's. The first reason for this resurgence was that Hopfield and Tank [38] designed a neural network to solve constraint satisfaction problems such as the "Traveling Salesman Problem", and Rumelhart, Hinton and Williams [86] refined and publicized an effective backpropagation algorithm for multi-layer neural networks which had been first investigated by Werbos [106]. The second reason was that some companies, most in Japan, had successfully made many fuzzy logic products such as fuzzy washing machines, fuzzy air conditioners and fuzzy subway trains [61]. With the rapid development for techniques of neural networks and fuzzy logic systems, neuro-fuzzy systems were attracting more and more interest since they may be more efficient and more powerful than either neural networks or fuzzy logic systems. Important progress with neuro-fuzzy systems has been made in recent years. In the theoretical realm, many effective learning algorithms for neuro-fuzzy systems have been developed and many structures of neuro-fuzzy systems have been proposed. For example, Jang's adaptive-network-based fuzzy inference systems, Lin's neural-network-based fuzzy logic control and decision system, Wang's adaptive fuzzy systems [103], the fuzzy ARTMAP by Carpenter, et al. [19], neural networks that learn from fuzzy if-then rules by Ishibuchi et al. [40], the fuzzy neural network learning fuzzy control rules and membership functions by fuzzy error backpropagation by Nauck and Kruse [76], etc. In application issues, neuro-fuzzy systems have been widely used in control systems, pattern recognition, consumer products, medicine, expert systems, fuzzy mathematics, game theory, etc. In the interdisciplinary aspect, more and more other techniques such as genetic algorithms, chaos, probability and AI are being applied to improve soft computing systems [1,2,7-10,15-25,28,31,32,34-36,39-46, 48,52,54,56-58,60,62,65,71-73,76-79,82,84,94,101,108,110,113-116]. Therefore, soft computing systems will become more and more powerful and efficient in the future.

With rapid development of theories and techniques for neural networks, fuzzy logic systems, genetic algorithms and the other new techniques, soft computing systems are attracting more and more interest in a variety of scientific and engineering fields. From an intelligent computation point of view, basic elements of intelligent computation in conventional neural networks are *numerical data* with high information granularity, which are used to model a system mapping numerical input data to numerical output data. Basic blocks of intelligent computation in pure fuzzy logic systems are *linguistic words* with low information granularity, which are used to build a system mapping linguistic input words to linguistic output words. In this context, there are two kinds of information systems which are (1) a system of computation with *numerical data* which come from natural environments and (2) a system of computation with *linguistic words* which exist in the human brain. In general, soft computing is the hybrid intelligent computation with both *numerical data* and *linguistic words*. Since neural networks and fuzzy systems have their own merits, hybrid neuro-fuzzy systems may be more powerful than either of them alone. From a system modeling point of view, a neuro-fuzzy system is a soft computing system with functional integration of neural networks and fuzzy logic, which is able to perform both *linguistic-words*-level fuzzy reasoning and *numerical-data*-level information processing. Thus a neuro-fuzzy system is capable of learning fuzzy IF-THEN rules from given data and processing numerical information from fuzzy knowledge bases, i.e., it has the ability of knowledge discovery and knowledge processing. In this sense, a neuro-fuzzy system is very useful since it can deal with uncertain and ill-defined information. Now the very important problem is how to build a powerful fuzzy reasoning engine for a neuro-fuzzy system in order to effectively perform complex nonlinear behaviors in real applications. In the aspect of logical reasoning, many existing fuzzy reasoning methods apply various fuzzy implications to try to make reasonable fuzzy decisions according to generalized modus ponens (GMP) and generalized modus tollens (GMT)[66,67,103,112,118]. Unfortunately, they are often not reasonable and not efficient. On the other hand, Takagi and Sugeno developed a fuzzy system by directly mapping input fuzzy sets to output crisp sets [93,95]. But Takagi and Sugeno didn't give a methodology to design a Takagi-Sugeno's fuzzy system directly based upon commonly used fuzzy rules such as IF (Temperature is *Very High*) and (Pressure is *Low*) THEN (Speed is *Very Low*). The ANFIS (Adaptive Neuro-Fuzzy Inference System) based on Takagi-Sugeno's fuzzy system developed by Jang is useful in many applications. However, since ANFIS still is not capable of extracting the commonly used fuzzy rules with both fuzzy IF and THEN parts, it's not convenient for us to acquire commonly used fuzzy knowledge from experts in a natural manner.

1.5 Main Issues

To solve the above problems, several theories and techniques have been developed in this book. The main contributions in the book are:

(1) Introducing a useful concept of a *primary* fuzzy set which is a fundamental granule of a fuzzy set. A primary-fuzzy-set-based fuzzy rule base contains more heuristic information than a non-primary-fuzzy-set-based one.

(2) Investigating a transformation from a fuzzy set to corresponding crisp features. Such a useful transformation increases the capability of a fuzzy neural network, in the sense that it can easily process linguistic words (i.e., fuzzy sets) by using crisp features. That is an important issue for CW because the fuzzy neural network provides us with a tool to map input words to corresponding output ones.

(3) Developing an effective fuzzy reasoning methodology based on primary fuzzy sets in order to generate reasonable and good outputs for a given fuzzy rule base. This new fuzzy reasoning methodology uses adaptive compensatory fuzzy operators rather than commonly used fixed operators such as Max and Min operators.

(4) Designing a universal fuzzy-reasoning-oriented fuzzy neural network which is capable of generating fuzzy IF-THEN rules from either numerical data or fuzzy sets, compressing a big fuzzy rule base to a small one and expanding a sparse fuzzy rule base into a non-sparse one. In addition, the new fuzzy neural network can easily be constructed directly from given fuzzy IF-THEN rules, by a heuristic learning algorithm, or by a genetic-guided training algorithm based on either fuzzy sets or numerical data.

(5) Constructing an efficient hybrid soft computing system incorporating fuzzy logic, neural network, genetic algorithms and AI techniques. Since all parameters in our new fuzzy neural network have known physical meanings (for example, centers and widths of Gaussian membership functions), they are easily initialized by either a heuristic method or genetic algorithms, and then a learning algorithm based on heuristically initialized parameters can have a better chance to get near-optimal (or optimal) values of parameters than that based on random ones. In addition, because the new fuzzy neural network has

a universal structure of CICO, CIFO, FICO and FIFO models, it can have stronger abilities in fuzzy knowledge discovery and fuzzy decision making than other existing systems.

(6) Applying the integrated soft computing system to several applications such as fuzzy control, compression of a fuzzy rule base, expansion of a sparse fuzzy rule base, fuzzy knowledge rediscovery from a fuzzy rule base, highly nonlinear system modeling, chaotic time series prediction, fuzzy games and pattern recognition.

Chapter 2

Fuzzy Compensation Principles

2.1 Fuzzy Yin-Yang Compensation

Figure 2.1. Taichi Yin(black)-Yang(white) compensation.

Before 4602 B.C., Taichi was created by the ancient Chinese [70]. Taichi consists of Yin Fish representing Yin and Yang Fish representing Yang. Yin (Yang) Fish has a Yang (Yin) Eye represented by a small circle in Figure 2.1. The S-type curve implies (1) Yin can be changed into Yang and Yang can also be changed into Yin; (2) the compensation of Yin and Yang (i.e., everything consists of two different parts which are Yin and Yang); and (3) the nonlinear relation of Yin and Yang. The principles of Taichi contain the principles of compensation of Yin and Yang. The Yang Eye in the Yin Fish means Yang is contained by Yin, and the Yin Eye in the Yang Fish means Yin is contained by Yang. In general, there exist a relative Yin containing Yang and a relative Yang containing Yin. The ratio between Yin and Yang dynamically changes and adapts. Taichi philosophically represents a dynamically optimal balance between Yin and Yang. Theoretically speaking, if we found an optimal balance (i.e., Taichi) between Yin and Yang, we could make an optimal decision between a pessimistic choice and an optimistic one. The reason is that if we only consider either a pessimistic decision or an optimistic one, we totally have two choices; if we analyze various cases between a pessimistic decision and an optimistic one, we can have infinite choices in which there may be many better decisions than either a pessimistic decision or an optimistic one. The compensation principle vividly represented by Taichi has widely been used in many real problems. For example, (1) the Boolean logic system uses 0 (Yin) and 1 (Yang) as basic logical values; (2) a fuzzy logic system uses a fuzzy false value in $[0, 0.5]$ and a fuzzy true value in $[0.5, 1.0]$ as basic fuzzy logical values. Intuitively, Taichi shown in Figure 2.1 is a perfect symbol of fuzzy logic.

The compensatory ANDs defined by Zimmermann and Zysno [117] also operate according to the principles of Yin-Yang compensation. Generally, we first need to consider both pessimistic situations (such as the worst case) (Yin) and optimistic situations (such as the best case) (Yang), and second, find a relatively good or even an optimal balance point (Taichi) between a pessimistic case and an optimistic case depending on different applications.

Based on the principles of compensation, we propose general operations of compensation as follows.

A general operation of compensation is described by a function such that

$$a \bigotimes b, \tag{2.1}$$

where $a \leq a \bigotimes b \leq b$, and $a \geq 0$ and $b > 0$ are a pessimistic value and an optimistic one, respectively, and $a < b$.

Usually, we choose simple compensatory operations for real applications because complex compensatory operations may make fuzzy reasoning too time-consuming and they also cannot guarantee to make a better decision than simple compensatory operations. We list two simple and useful compensatory operations as follows.

Example 1: A linear compensatory operation is given by

$$a \bigotimes b = (1 - \gamma)a + \gamma b, \tag{2.2}$$

where $0 \leq \gamma \leq 1$ and $0 \leq a < b$.

Example 2: A nonlinear compensatory operation is given by

$$a \bigotimes b = a^{1-\gamma} b^{\gamma}, \tag{2.3}$$

where $0 \leq \gamma \leq 1$, $0 < a < b$.

Interestingly, commonly used fuzzy operations Min and Max are just special cases of the compensatory operations in the above examples such that

$$Min(a,b) = a \bigotimes b|_{\gamma=0} = a, \tag{2.4}$$

$$Max(a,b) = a \bigotimes b|_{\gamma=1} = b. \tag{2.5}$$

Therefore, conventionally used fixed Min and Max fuzzy operations are two extreme cases of the given compensatory operations. Because we can adjust the value of γ, we are able to adaptively optimize a compensatory operation for different applications. In this sense, a compensatory operation is more adaptive than a fixed fuzzy operation such as Min or Max.

2.2 Compensation of Fuzzy CNF and Fuzzy DNF

Because the 2-variable-8-entry Kaufmann's fuzzy truth table which contains all possible orders of a, $\bar{a}$, b and $\bar{b}$ for $a, b \in [0, 1]$, $\bar{a} = 1 - a$ and $\bar{b} = 1 - b$ [53], it is valid for Zadeh's fuzzy logic using Max & Min operations. For the general t-norm-t-conorm cases, Kaufmann's fuzzy truth tables are no longer useful because they don't contain such one-literal-based phrases (clauses) as $a \cap \bar{a}$, $a \cap a$, $a \cup \bar{a}$, and $a \cup a$, etc. To solve this problem, AND_n^m map and OR_n^m map were developed to contain all possible logical sub-concepts (i.e. phrases and clauses) [50,51]. As a result, AND_n^m map (OR_n^m map), a generalized framework of fuzzy map [88], is a complete space on which axioms-based CNFs and DNFs are built. Therefore, we use AND_n^m map to generate axioms-based CNFs and DNFs so as to avoid losing meaningful sub-concepts.

Excluded middle law ($a \sqcup \bar{a} = 1$) and contradiction law ($a \sqcap \bar{a} = 0$) for $a \in \{0, 1\}$ are very useful in Boolean logic, but they are no longer true for Zadeh's fuzzy logic using Max ($\vee$) & Min ($\wedge$) operators (i.e., $a \wedge \bar{a} \neq 1$ and $a \vee \bar{a} \neq 0$ for $a \in [0, 1]$) and the other general fuzzy logics using t-norm ($\triangle$) & t-conorm ($\triangledown$) operators (i.e., $a \triangledown \bar{a} \neq 1$ and $a \triangle \bar{a} \neq 0$ for $a \in [0, 1]$). Turksen [97-99] used the fuzzy normal forms to represent $a \triangledown \bar{a}$ and $a \triangle \bar{a}$, and analyzed fuzzy CNF-DNF-based logical structures of them. The fuzzy normal forms of $a \triangledown \bar{a}$ and $a \triangle \bar{a}$ are constructed by substituting $\bar{a}$ in place of b in the CNF and DNF expressions of

$a \triangledown b$ and $a \triangle b$ respectively based on the 2-variable-8-entry fuzzy truth table which contains all possible orders of a, $\bar{a}$, b and $\bar{b}$ for $a, b \in [0, 1]$, $\bar{a} = 1 - a$ and $\bar{b} = 1 - b$ [99]. Actually, $a \triangledown \bar{a}$ and $a \triangle \bar{a}$ are 1-variable logical expressions, so we have to use 1-variable fuzzy truth tables to construct 1-variable axioms-based CNFs and DNFs. Importantly, because AND_n^m map and OR_n^m map contain all possible logical sub-concepts (i.e. phrases and clauses) [1], AND_n^m map (OR_n^m map), a generalized framework of fuzzy map, is a complete space on which axioms-based CNFs and DNFs are built. Therefore, we use AND_n^m map to generate CNFs and DNFs of $a \triangledown \bar{a}$ and $a \triangle \bar{a}$ so as to avoid losing meaningful sub-concepts.

2.2.1 Boolean Truth Table and Karnaugh Map

Suppose crisp sets' membership values $a, b \in \{0, 1\}$, $\bar{a} = 1 - a$ and $\bar{b} = 1 - b$. A and $\bar{A}$ are the degrees of truth of an affirmative statement and a negative statement, respectively, and B and $\bar{B}$ are the degrees of truth of another affirmative statement and another negative statement, respectively. Importantly, $A, \bar{A}, B, \bar{B} \in \{0, 1\}$, i.e., Boolean truth values are binary (100% false or 100% true). $\cap$ and $\cup$ are the logical AND and OR for degrees of truth respectively, $\sqcap$ and $\sqcup$ are the computational Boolean AND and OR for degrees of membership, respectively.

Table 2.1: 2-variable-2-dimensional Boolean Truth Table.

A	B	$(A \sqcap B) \sqcup (\bar{A} \sqcap B)$	Phrases
0	0	0	$\bar{a} \sqcap b$
0	1	1	$\bar{a} \sqcap b$
1	0	1	$a \sqcap b$
1	1	0	$a \sqcap b$

Since a Boolean truth table is logically equivalent to a Karnaugh map, both methods generate the same results. For example, by using 2-variable-2-dimensional Boolean truth table (Table 2.1) and 2-variable-2-dimensional Karnaugh map (Table 2.2), we can easily get the 2-variable-2-dimensional CNF and DNF for $(a \sqcap \bar{b}) \sqcup (\bar{a} \sqcap b)$ as follows,

$$CNF_2^2[(a \sqcap \bar{b}) \sqcup (\bar{a} \sqcap b)] = (a \sqcup b) \sqcap (\bar{a} \sqcup \bar{b}), \qquad (2.6)$$

$$DNF_2^2[(a \sqcap \bar{b}) \sqcup (\bar{a} \sqcap b)] = (a \sqcap \bar{b}) \sqcup (\bar{a} \sqcap b). \qquad (2.7)$$

Since both Boolean truth table and Karnaugh map can completely express all meaningful sub-concepts of Boolean logic, we can use them to generate CNFs and DNFs without losing any useful information. Therefore, both Boolean truth table and Karnaugh map are of conceptual completeness and logical consistency.

Table 2.2: 2-variable-2-dimensional Karnaugh Map.

	$\bar{b}$	b
$\bar{a}$	0	1
a	1	0

2.2.2 Kaufmann's Fuzzy Truth Table and Fuzzy Map

Suppose fuzzy sets' membership values $a, b \in \{0, 1\}$, $\bar{a} = 1 - a$ and $\bar{b} = 1 - b$. A and $\bar{A}$ are the degrees of truth of an affirmative statement and a negative statement, respectively, and B and $\bar{B}$ are the degrees of truth of another affirmative statement and another negative statement, respectively. Very importantly, $A, \bar{A}, B, \bar{B} \in [0, 1]$, i.e., fuzzy truth values are degrees of truth (such as 75% true, 25% false). $\cap$ and $\cup$ are the logical AND and OR for degrees of truth, respectively, $\triangle$ and $\triangledown$ are the computational AND (i.e. t-norm) and OR (i.e. t-conorm) for degrees of membership, respectively. A Boolean truth table is logically equivalent to a Karnaugh map. Similarly, we can create a corresponding Kaufmann's fuzzy map, which is logically equivalent to Kaufmann's fuzzy truth table. At first, let's examine the applicability of Kaufmann's fuzzy truth table for all kinds of t-norms and t-conorms with the following 3 examples.
(1) Generating CNF and DNF of $a \triangle b$

Table 2.3: Kaufmann's Fuzzy Map for $A \cap B$.

	$\bar{b}$	b			$\bar{a}$	a
$\bar{a}$	≤ 0.5	≤ 0.5		$\bar{b}$	≤ 0.5	≤ 0.5
a	≤ 0.5	≥ 0.5		b	≤ 0.5	≥ 0.5

In Table 2.4, $F(A, B) = A \cap B$. According to Table 2.3 or Table 2.4, we have

$$CNF_2^2[a \triangle b] = (a \triangledown b) \triangle (a \triangledown \bar{b}) \triangle (\bar{a} \triangledown b) \triangle$$
$$(b \triangledown a) \triangle (b \triangledown \bar{a}) \triangle (\bar{b} \triangledown a), \quad (2.8)$$
$$DNF_2^2[a \triangle b] = (a \triangle b) \triangledown (b \triangle a). \quad (2.9)$$

Table 2.4: Kaufmann's Fuzzy Truth Table for $A \cap B$.

A	B	$F(A,B)$	Phrases	B	A	$F(A,B)$	Phrases
$[0, 0.5]$	$[0, 0.5]$	≤ 0.5	$\bar{a} \triangle \bar{b}$	$[0, 0.5]$	$[0, 0.5]$	≤ 0.5	$\bar{b} \triangle \bar{a}$
$[0, 0.5]$	$[0.5, 1]$	≤ 0.5	$\bar{a} \triangle b$	$[0, 0.5]$	$[0.5, 1]$	≤ 0.5	$\bar{b} \triangle a$
$[0.5, 1]$	$[0, 0.5]$	≤ 0.5	$a \triangle \bar{b}$	$[0.5, 1]$	$[0, 0.5]$	≤ 0.5	$b \triangle \bar{a}$
$[0.5, 1]$	$[0.5, 1]$	≥ 0.5	$a \triangle b$	$[0.5, 1]$	$[0.5, 1]$	≥ 0.5	$b \triangle a$

(2) Generating CNF and DNF of $(a \triangle \bar{a}) \triangledown (a \triangle b)$

Table 2.5: Kaufmann's Fuzzy Map for $(A \cap \bar{A}) \cup (A \cap B)$.

	b	b			$\bar{a}$	a
$\bar{a}$	≤ 0.5	≤ 0.5		b	≤ 0.5	≤ 0.5
a	≤ 0.5	≥ 0.5		b	≤ 0.5	≥ 0.5

Table 2.6: Kaufmann's Fuzzy Truth Table for $(A \cap \bar{A}) \cup (A \cap B)$.

A	B	$F(A,B)$	Phrases	B	A	$F(A,B)$	Phrases
$[0, 0.5]$	$[0, 0.5]$	≤ 0.5	$\bar{a} \triangle \bar{b}$	$[0, 0.5]$	$[0, 0.5]$	≤ 0.5	$\bar{b} \triangle \bar{a}$
$[0, 0.5]$	$[0.5, 1]$	≤ 0.5	$\bar{a} \triangle b$	$[0, 0.5]$	$[0.5, 1]$	≤ 0.5	$\bar{b} \triangle a$
$[0.5, 1]$	$[0, 0.5]$	≤ 0.5	$a \triangle \bar{b}$	$[0.5, 1]$	$[0, 0.5]$	≤ 0.5	$b \triangle \bar{a}$
$[0.5, 1]$	$[0.5, 1]$	≥ 0.5	$a \triangle b$	$[0.5, 1]$	$[0.5, 1]$	≥ 0.5	$b \triangle a$

In Table 2.6, $F(A,B) = (A \cap \bar{A}) \cup (A \cap B)$. According to Table 2.5 or Table 2.6, we have

$$CNF_2^2[(a \triangle \bar{a}) \triangledown (a \triangle b)] = (a \triangledown b) \triangle (a \triangledown \bar{b}) \triangle (\bar{a} \triangledown b) \triangle$$
$$(b \triangledown a) \triangle (b \triangledown \bar{a}) \triangle (\bar{b} \triangledown a), \quad (2.10)$$
$$DNF_2^2[(a \triangle \bar{a}) \triangledown (a \triangle b)] = (a \triangle b) \triangledown (b \triangle a). \quad (2.11)$$

(3) Generating CNF and DNF of $(a \triangle \bar{a}) \triangledown (b \triangle \bar{b})$

In Table 2.8, $F(A,B) = (A \cap \bar{A}) \cup (B \cap \bar{B})$. According to Table 2.7 or Table 2.8, we have:

$$CNF_2^2[(a \triangle \bar{a}) \triangledown (b \triangle \bar{b})] = (a \triangledown b) \triangle (a \triangledown \bar{b}) \triangle (\bar{a} \triangledown b) \triangle (\bar{a} \triangledown \bar{b}) \triangle$$
$$(b \triangledown a) \triangle (b \triangledown \bar{a}) \triangle (\bar{b} \triangledown a) \triangle (\bar{b} \triangledown \bar{a}) (2.12)$$
$$DNF_2^2[(a \triangle \bar{a}) \triangledown (b \triangle \bar{b})] = ?. \quad (2.13)$$

Sec. 2.2. Compensation of Fuzzy CNF and Fuzzy DNF

Table 2.7: Kaufmann's Fuzzy Map for $(A \cap \bar{A}) \cup (B \cap \bar{B})$.

	b	b			$\bar{a}$	a
$\bar{a}$	≤ 0.5	≤ 0.5		b	≤ 0.5	≤ 0.5
a	≤ 0.5	≤ 0.5		b	≤ 0.5	≤ 0.5

Table 2.8: Kaufmann's Fuzzy Truth Table for $(A \cap \bar{A}) \cup (B \cap \bar{B})$.

A	B	$F(A,B)$	Phrases		B	A	$F(A,B)$	Phrases
$[0, 0.5]$	$[0, 0.5]$	≤ 0.5	$\bar{a} \triangle \bar{b}$		$[0, 0.5]$	$[0, 0.5]$	≤ 0.5	$\bar{b} \triangle \bar{a}$
$[0, 0.5]$	$[0.5, 1]$	≤ 0.5	$\bar{a} \triangle b$		$[0, 0.5]$	$[0.5, 1]$	≤ 0.5	$\bar{b} \triangle a$
$[0.5, 1]$	$[0, 0.5]$	≤ 0.5	$a \triangle \bar{b}$		$[0.5, 1]$	$[0, 0.5]$	≤ 0.5	$b \triangle \bar{a}$
$[0.5, 1]$	$[0.5, 1]$	≤ 0.5	$a \triangle b$		$[0.5, 1]$	$[0.5, 1]$	≤ 0.5	$b \triangle a$

By analyzing the above CNF and DNF expressions, we can discover some interesting but unreasonable phenomena as follows,
(a) $CNF_2^2[a \triangle b]$ is the same as $CNF_2^2[(a \triangle a) \triangledown (a \triangle b)]$, and $DNF_2^2[a \triangle b]$ is the same as $DNF_2^2[(a \triangle a) \triangledown (a \triangle b)]$. The results indicate that the sub-concept $a \triangle a$ is no longer useful and valid. The reason resulting in such unreasonable phenomena is that Kaufmann's fuzzy truth table only contains 8 sub-concepts and misses the other 8 meaningful sub-concepts such as $a \triangle a$, $a \triangle \bar{a}$, $b \triangle b$ and $b \triangle \bar{b}$.
(b) Since $CNF_2^2[(a \triangle \bar{a}) \triangledown (b \triangle \bar{b})]$ contains all 8 clauses which are derived from corresponding 8 phrases in Kaufmann's fuzzy truth table (Table 2.8), a difficult problem is "$DNF_2^2[(a \triangle \bar{a}) \triangledown (b \triangle \bar{b})] = ?$". Intuitively, a reasonable result is $DNF_1^2[(a \triangle \bar{a}) \triangledown (b \triangle \bar{b})] = 0$. But, the original concept $(a \triangle \bar{a}) \triangledown (b \triangle \bar{b})$ is just a meaningful DNF expression, thus $DNF_1^2[(a \triangle \bar{a}) \triangledown (a \triangle b)]$ must contain at least $(a \triangle \bar{a}) \triangledown (b \triangle \bar{b})$. Obviously, Kaufmann's fuzzy truth table doesn't contain all possible sub-concepts on which CNFs and DNFs are constructed. In other words, Kaufmann's fuzzy truth tables lose many meaningful sub-concepts which are useful to represent some sub-concepts for t-norms and t-conorms.

At this point, an interesting question is "what is a universal fuzzy truth table which contains all possible sub-concepts on which CNFs and DNFs are constructed?". The next section will answer this important question.

2.2.3 Universal Fuzzy Truth Table and AND_n^m Map

Suppose fuzzy sets' membership values $a, b \in [0, 1]$, $\bar{a} = 1 - a$ and $\bar{b} = 1 - b$. A and $\bar{A}$ are the degrees of truth of an affirmative statement and a negative statement, respectively, and B and $\bar{B}$ are the degrees of truth of another affirmative statement and another negative statement, respectively. $A, \bar{A}, B, \bar{B} \in [0, 1]$. $\cap$ and $\cup$ are the logical AND and OR for degrees of truth, respectively, $\triangle$ and $\triangledown$ are the computational AND (i.e. t-norm) and OR (i.e. t-conorm) for degrees of membership, respectively. Since AND_n^m map completely describing all possible sub-concepts is logically equivalent to the corresponding n-variable-m-dimensional truth table [1], both methods can achieve the same results.

The algorithm for generating a CNF and a DNF of a 2-variable-2-dimensional logical expression $F(A, B)$ by a 2-variable-2-dimensional universal fuzzy truth table is given below,

BEGIN
Step 1: Create a 2-variable-2-dimensional universal fuzzy truth table with 16 entries.
Step 2: Calculate the values of $F(A, B)$ for 16 entries:
If $0 \leq F(A, B) \leq 0.5$

Then write ≤ 0.5 into the corresponding entry on the column $F(A, B)$;
If $0.5 \leq F(A, B) \leq 1$

Then write ≥ 0.5 into the corresponding entry on the column $F(A, B)$;
If $F(A, B) = 0.5$

Then write 0.5 into the corresponding entry on the column $F(A, B)$;
If $F(A, B)$ is uncertain

Then write ? into the corresponding entry on the column $F(A, B)$.
Step 3: Construct the DNF of $F(A, B)$:

Combine the phrases with $F(A, B) \geq 0.5$ or $F(A, B) = 0.5$ with disjunctions.
Step 4: Construct the CNF of $F(A, B)$:

Make negations of the phrases with $F(A, B) \leq 0.5$ or $F(A, B) = 0.5$ to get corresponding clauses by using DeMorgan's Laws and the Involution Law, then combine the resulting clauses with conjunctions.
Step 5: End (? entries are invalid in constructing a CNF and a DNF).
END.

Similarly, we can use the above algorithm to generate a CNF and a DNF of a 2-variable-2-dimensional logical expression $F(A, B)$ by an AND_2^2 map.

For the unsolved problems in Section 2.2.2, we use both methods to generate meaningful and complete CNFs and DNFs for $a \triangledown b$, $(a \triangledown \bar{a}) \triangle (a \triangledown b)$ and $(a \triangle \bar{a}) \triangle (b \triangle \bar{b})$.

Sec. 2.2. Compensation of Fuzzy CNF and Fuzzy DNF

(1) Generating CNF and DNF of $a \triangle b$

Table 2.9: AND_2^2 Map for $A \cap B$.

	$\bar{a}$	a	b	$\bar{b}$
$\bar{a}$	≤ 0.5	≤ 0.5	≤ 0.5	≤ 0.5
a	≤ 0.5	?	≤ 0.5	≥ 0.5
b	≤ 0.5	≤ 0.5	≤ 0.5	≤ 0.5
$\bar{b}$	≤ 0.5	≥ 0.5	≤ 0.5	?

In Table 2.10, $F(A, B) = A \cap B$. According to Table 2.9 or 2.10, we have

$$\begin{aligned}
CNF_2^2[a \triangle b] &= (a \triangledown a) \triangle (a \triangledown \bar{a}) \triangle (\bar{a} \triangledown a) \triangle (a \triangledown b) \\
&\triangle (a \triangledown \bar{b}) \triangle (\bar{a} \triangledown b) \triangle (b \triangledown a) \triangle (b \triangledown \bar{a}) \\
&\triangle (\bar{b} \triangledown a) \triangle (b \triangledown b) \triangle (b \triangledown \bar{b}) \triangle (\bar{b} \triangledown b), \quad (2.14)\\
DNF_2^2[a \triangle b] &= (a \triangle b) \triangledown (b \triangle a). \quad (2.15)
\end{aligned}$$

Table 2.10: 2-variable-2-dimensional Universal Fuzzy Truth Table for $A \cap B$.

A	A	$F(A,B)$	Phrases	A	B	$F(A,B)$	Phrases
$[0, 0.5]$	$[0, 0.5]$	≤ 0.5	$\bar{a} \triangle \bar{a}$	$[0, 0.5]$	$[0, 0.5]$	≤ 0.5	$\bar{a} \triangle \bar{b}$
$[0, 0.5]$	$[0.5, 1]$	≤ 0.5	$\bar{a} \triangle a$	$[0, 0.5]$	$[0.5, 1]$	≤ 0.5	$\bar{a} \triangle b$
$[0.5, 1]$	$[0, 0.5]$	≤ 0.5	$a \triangle \bar{a}$	$[0.5, 1]$	$[0, 0.5]$	≤ 0.5	$a \triangle \bar{b}$
$[0.5, 1]$	$[0.5, 1]$	?	$a \triangle a$	$[0.5, 1]$	$[0.5, 1]$	≥ 0.5	$a \triangle b$
B	A	$F(A,B)$	Phrases	B	B	$F(A,B)$	Phrases
$[0, 0.5]$	$[0, 0.5]$	≤ 0.5	$\bar{b} \triangle \bar{a}$	$[0, 0.5]$	$[0, 0.5]$	≤ 0.5	$\bar{b} \triangle \bar{b}$
$[0, 0.5]$	$[0.5, 1]$	≤ 0.5	$\bar{b} \triangle a$	$[0, 0.5]$	$[0.5, 1]$	≤ 0.5	$\bar{b} \triangle b$
$[0.5, 1]$	$[0, 0.5]$	≤ 0.5	$b \triangle \bar{a}$	$[0.5, 1]$	$[0, 0.5]$	≤ 0.5	$b \triangle \bar{b}$
$[0.5, 1]$	$[0.5, 1]$	≥ 0.5	$b \triangle a$	$[0.5, 1]$	$[0.5, 1]$	?	$b \triangle b$

(2) Generating CNF and DNF of $(a \triangle \bar{a}) \triangledown (a \triangle b)$

In Table 2.12, $F(A, B) = (A \cap \bar{A}) \cup (A \cap B)$. According to Table 2.11 or 2.12, we have

$$\begin{aligned}
CNF_2^2[(a \triangle a) \triangledown (a \triangle b)] &= (a \triangledown a) \triangle (a \triangledown \bar{a}) \triangle (\bar{a} \triangledown a) \triangle \\
&(a \triangledown b) \triangle (a \triangledown \bar{b}) \triangle (\bar{a} \triangledown b) \triangle \\
&(b \triangledown a) \triangle (b \triangledown \bar{a}) \triangle (\bar{b} \triangledown a) \triangle \\
&(b \triangledown b) \triangle (b \triangledown \bar{b}) \triangle (\bar{b} \triangledown b), \quad (2.16)\\
DNF_2^2[(a \triangle \bar{a}) \triangledown (a \triangle b)] &= (a \triangle \bar{a}) \triangledown (\bar{a} \triangle a) \triangledown \\
&(a \triangle b) \triangledown (b \triangle a). \quad (2.17)
\end{aligned}$$

Table 2.11: AND_2^2 Map for $(A \cap \bar{A}) \cup (A \cap B)$.

	$\bar{a}$	a	b	$\bar{b}$
$\bar{a}$	≤ 0.5	0.5	≤ 0.5	≤ 0.5
a	0.5	?	≤ 0.5	≥ 0.5
b	≤ 0.5	≤ 0.5	≤ 0.5	≤ 0.5
$\bar{b}$	≤ 0.5	≥ 0.5	≤ 0.5	?

Table 2.12: 2-variable-2-dimensional Universal Fuzzy Truth Table for $(A \cap \bar{A}) \cup (A \cap B)$.

A	A	$F(A,B)$	Phrases	A	B	$F(A,B)$	Phrases
$[0,0.5]$	$[0,0.5]$	≤ 0.5	$\bar{a} \triangle \bar{a}$	$[0,0.5]$	$[0,0.5]$	≤ 0.5	$\bar{a} \triangle \bar{b}$
$[0,0.5]$	$[0.5,1]$	0.5	$\bar{a} \triangle a$	$[0,0.5]$	$[0.5,1]$	≤ 0.5	$\bar{a} \triangle b$
$[0.5,1]$	$[0,0.5]$	0.5	$a \triangle \bar{a}$	$[0.5,1]$	$[0,0.5]$	≤ 0.5	$a \triangle \bar{b}$
$[0.5,1]$	$[0.5,1]$	?	$a \triangle a$	$[0.5,1]$	$[0.5,1]$	≥ 0.5	$a \triangle b$
B	A	$F(A,B)$	Phrases	B	B	$F(A,B)$	Phrases
$[0,0.5]$	$[0,0.5]$	≤ 0.5	$\bar{b} \triangle \bar{a}$	$[0,0.5]$	$[0,0.5]$	≤ 0.5	$\bar{b} \triangle \bar{b}$
$[0,0.5]$	$[0.5,1]$	≤ 0.5	$\bar{b} \triangle a$	$[0,0.5]$	$[0.5,1]$	≤ 0.5	$\bar{b} \triangle b$
$[0.5,1]$	$[0,0.5]$	≤ 0.5	$b \triangle \bar{a}$	$[0.5,1]$	$[0,0.5]$	≤ 0.5	$b \triangle \bar{b}$
$[0.5,1]$	$[0.5,1]$	≥ 0.5	$b \triangle a$	$[0.5,1]$	$[0.5,1]$	?	$b \triangle b$

(3) Generating CNF and DNF of $(a \triangle \bar{a}) \triangledown (b \triangle \bar{b})$

Table 2.13: AND_2^2 Map for $(A \cap \bar{A}) \cup (B \cap \bar{B})$.

	$\bar{a}$	a	b	$\bar{b}$
$\bar{a}$	≤ 0.5	0.5	≤ 0.5	≤ 0.5
a	0.5	≤ 0.5	≤ 0.5	≤ 0.5
b	≤ 0.5	≤ 0.5	≤ 0.5	0.5
$\bar{b}$	≤ 0.5	≤ 0.5	0.5	≤ 0.5

In Table 2.14, $F(A,B) = (A \cap \bar{A}) \cup (B \cap \bar{B})$. According to Table 2.13 or Table 2.14, we have

$$DNF_2^2[(a \triangle \bar{a}) \triangledown (b \triangle \bar{b})] = (a \triangle \bar{a}) \triangledown (\bar{a} \triangle a)$$
$$\triangledown (b \triangle \bar{b}) \triangledown (\bar{b} \triangle b), \qquad (2.18)$$

$$CNF_2^2[(a \triangle \bar{a}) \triangledown (b \triangle \bar{b})] = (a \triangledown a) \triangle (a \triangledown \bar{a}) \triangle (\bar{a} \triangledown a) \triangle (\bar{a} \triangledown \bar{a})$$
$$\triangle (a \triangledown b) \triangle (a \triangledown \bar{b}) \triangle (\bar{a} \triangledown b) \triangle$$
$$(\bar{a} \triangledown \bar{b}) \triangle (b \triangledown a) \triangle (b \triangledown \bar{a}) \triangle$$
$$(\bar{b} \triangledown a) \triangle (\bar{b} \triangledown \bar{a}) \triangle (b \triangledown b) \triangle$$
$$(b \triangledown \bar{b}) \triangle (\bar{b} \triangledown b) \triangle (\bar{b} \triangledown \bar{b}). \qquad (2.19)$$

Table 2.14: 2-variable-2-dimensional Universal Fuzzy Truth Table for $(A \cap \bar{A}) \cup (B \cap \bar{B})$.

A	A	$F(A,B)$	Phrases	A	B	$F(A,B)$	Phrases
$[0, 0.5]$	$[0, 0.5]$	≤ 0.5	$\bar{a} \triangle \bar{a}$	$[0, 0.5]$	$[0, 0.5]$	≤ 0.5	$\bar{a} \triangle \bar{b}$
$[0, 0.5]$	$[0.5, 1]$	0.5	$\bar{a} \triangle a$	$[0, 0.5]$	$[0.5, 1]$	≤ 0.5	$\bar{a} \triangle b$
$[0.5, 1]$	$[0, 0.5]$	0.5	$a \triangle \bar{a}$	$[0.5, 1]$	$[0, 0.5]$	≤ 0.5	$a \triangle \bar{b}$
$[0.5, 1]$	$[0.5, 1]$	≤ 0.5	$a \triangle a$	$[0.5, 1]$	$[0.5, 1]$	≤ 0.5	$a \triangle b$
B	A	$F(A,B)$	Phrases	B	B	$F(A,B)$	Phrases
$[0, 0.5]$	$[0, 0.5]$	≤ 0.5	$b \triangle \bar{a}$	$[0, 0.5]$	$[0, 0.5]$	≤ 0.5	$b \triangle \bar{b}$
$[0, 0.5]$	$[0.5, 1]$	≤ 0.5	$b \triangle a$	$[0, 0.5]$	$[0.5, 1]$	0.5	$b \triangle \bar{b}$
$[0.5, 1]$	$[0, 0.5]$	≤ 0.5	$b \triangle \bar{a}$	$[0.5, 1]$	$[0, 0.5]$	0.5	$b \triangle \bar{b}$
$[0.5, 1]$	$[0.5, 1]$	≤ 0.5	$b \triangle a$	$[0.5, 1]$	$[0.5, 1]$	≤ 0.5	$b \triangle b$

2.3 2-variable-2-dimensional CNFs and DNFs

CNF_2^2 and DNF_2^2 expressions of the 16 combined concepts were studied in [50,99]. Here, we discuss how to use 2-variable-2-dimensional universal fuzzy truth tables to generate CNF_2^2 and DNF_2^2 expressions of the 16 combined concepts when there exist DeMorgan's Laws, the involution law, commutativity laws and associativity laws. Because we have shown how to use 2-variable-2-dimensional universal fuzzy truth tables to generate CNF_2^2 and DNF_2^2 expressions in the above section, we only list all typical results for simplicity. For convenience, suppose $x \in \{\bar{a}, a\}$ and $y \in \{\bar{b}, b\}$.

(1) Affirmative 1:

$$DNF_2^2[1] = (a \triangle a) \triangledown (a \triangle \bar{a}) \triangledown (\bar{a} \triangle a) \triangledown (\bar{a} \triangle \bar{a}) \triangledown$$
$$(a \triangle b) \triangledown (a \triangle \bar{b}) \triangledown (\bar{a} \triangle b) \triangledown (\bar{a} \triangle \bar{b}) \triangledown$$
$$(b \triangle a) \triangledown (b \triangle \bar{a}) \triangledown (\bar{b} \triangle a) \triangledown (\bar{b} \triangle \bar{a}) \triangledown$$
$$(b \triangle b) \triangledown (b \triangle \bar{b}) \triangledown (\bar{b} \triangle b) \triangledown (\bar{b} \triangle \bar{b}), \qquad (2.20)$$
$$CNF_2^2[1] = 1. \qquad (2.21)$$

(2) Negative 0:

$$\begin{aligned}CNF_2^2[0] &= (a \triangledown a) \triangle (a \triangledown \bar{a}) \triangle (\bar{a} \triangledown a) \triangle (\bar{a} \triangledown \bar{a}) \triangle \\ &\quad (a \triangledown b) \triangle (a \triangledown \bar{b}) \triangle (\bar{a} \triangledown b) \triangle (\bar{a} \triangledown \bar{b}) \triangle \\ &\quad (b \triangledown a) \triangle (b \triangledown \bar{a}) \triangle (\bar{b} \triangledown a) \triangle (\bar{b} \triangledown \bar{a}) \triangle \\ &\quad (b \triangledown b) \triangle (b \triangledown \bar{b}) \triangle (\bar{b} \triangledown b) \triangle (\bar{b} \triangledown \bar{b}), \quad (2.22)\\ DNF_2^2[0] &= 0. \quad (2.23)\end{aligned}$$

(3) $(a \triangle b) \triangledown (\bar{a} \triangle \bar{b})$:

$$\begin{aligned}CNF_2^2[(a \triangle b) \triangledown (\bar{a} \triangle \bar{b})] &= (a \triangledown \bar{a}) \triangle (\bar{a} \triangledown a) \triangle (a \triangledown \bar{b}) \triangle (\bar{a} \triangledown b) \triangle \\ &\quad (b \triangledown \bar{a}) \triangle (\bar{b} \triangledown a) \triangle (b \triangledown \bar{b}) \triangle (\bar{b} \triangledown b), (2.24)\\ DNF_2^2[(a \triangle b) \triangledown (\bar{a} \triangle \bar{b})] &= (a \triangle \bar{a}) \triangledown (\bar{a} \triangle a) \triangledown (a \triangle b) \triangledown (\bar{a} \triangle \bar{b}) \\ &\quad (b \triangle a) \triangledown (\bar{b} \triangle \bar{a}) \triangledown (b \triangle \bar{b}) \triangledown (\bar{b} \triangle b). (2.25)\end{aligned}$$

(4) $(a \triangle \bar{b}) \triangledown (\bar{a} \triangle b)$:

$$\begin{aligned}CNF_2^2[(a \triangle \bar{b}) \triangledown (\bar{a} \triangle b)] &= (a \triangledown \bar{a}) \triangle (\bar{a} \triangledown a) \triangle (a \triangledown b) \triangle (\bar{a} \triangledown \bar{b}) \triangle \\ &\quad (b \triangledown a) \triangle (\bar{b} \triangledown \bar{a}) \triangle (b \triangledown \bar{b}) \triangle (\bar{b} \triangledown b), (2.26)\\ DNF_2^2[(a \triangle \bar{b}) \triangledown (\bar{a} \triangle b)] &= (a \triangle \bar{a}) \triangledown (\bar{a} \triangle a) \triangledown (\bar{a} \triangle b) \triangledown (a \triangle \bar{b}) \\ &\quad (\bar{b} \triangle a) \triangledown (\bar{b} \triangle a) \triangledown (b \triangle \bar{b}) \triangledown (\bar{b} \triangle b). (2.27)\end{aligned}$$

(5) $x \triangle y$ type:

$$\begin{aligned}CNF_2^2[x \triangle y] &= (x \triangledown x) \triangle (x \triangledown \bar{x}) \triangle (\bar{x} \triangledown x) \triangle (x \triangledown y) \\ &\quad \triangle (x \triangledown \bar{y}) \triangle (\bar{x} \triangledown y) \triangle (b \triangledown x) \triangle (y \triangledown \bar{x}) \\ &\quad \triangle (\bar{y} \triangledown x) \triangle (y \triangledown y) \triangle (y \triangledown \bar{y}) \triangle (\bar{y} \triangledown y), \quad (2.28)\\ DNF_2^2[x \triangle y] &= (x \triangle y) \triangledown (y \triangle x). \quad (2.29)\end{aligned}$$

(6) $x \triangledown y$ type:

$$\begin{aligned}CNF_2^2[x \triangledown y] &= (x \triangledown y) \triangle (y \triangledown x), \quad (2.30)\\ DNF_2^2[x \triangledown y] &= (x \triangle x) \triangledown (x \triangle \bar{x}) \triangledown (\bar{x} \triangle x) \triangledown (x \triangle y) \triangledown \\ &\quad (x \triangle \bar{y}) \triangledown (\bar{x} \triangle y) \triangledown (y \triangle x) \triangledown (y \triangle \bar{x}) \triangledown \\ &\quad (\bar{y} \triangle x) \triangledown (y \triangle y) \triangledown (y \triangle \bar{y}) \triangledown (\bar{y} \triangle y). \quad (2.31)\end{aligned}$$

(7) x type:

$$\begin{aligned}CNF_2^2[x] &= (x \triangledown x) \triangle (x \triangledown \bar{x}) \triangle (\bar{x} \triangledown x) \triangle (x \triangledown y) \\ &\quad \triangle (x \triangledown \bar{y}) \triangle (y \triangledown x) \triangle (\bar{y} \triangledown x), \quad (2.32)\\ DNF_2^2[x] &= (x \triangle x) \triangledown (x \triangle \bar{x}) \triangledown (\bar{x} \triangle x) \triangledown (x \triangle y) \triangledown \\ &\quad (x \triangle \bar{y}) \triangledown (\bar{y} \triangle x) \triangledown (y \triangle x). \quad (2.33)\end{aligned}$$

(8) y type:

$$\begin{aligned}
CNF_2^2[y] &= (y \triangledown y) \triangle (y \triangledown \bar{y}) \triangle (\bar{y} \triangledown y) \triangle (y \triangledown x) \\
&\quad \triangle (y \triangledown \bar{x}) \triangle (x \triangledown y) \triangle (\bar{x} \triangledown y), \quad\quad (2.34) \\
DNF_2^2[y] &= (y \triangle y) \triangledown (y \triangle \bar{y}) \triangledown (\bar{y} \triangle y) \triangledown (y \triangle x) \triangledown \\
&\quad (y \triangle \bar{x}) \triangledown (\bar{x} \triangle y) \triangledown (x \triangle y). \quad\quad (2.35)
\end{aligned}$$

2.4 2-variable-m-dimensional CNFs and DNFs for $m = 3, 4$

In real applications, we often meet not only 2-dimensional logical expressions as shown in the above sections but also m-dimensional logical expressions for $m \geq 3$. For example, we have a 3-dimensional logical expression and a 4-dimensional logical expression as follows,

(1) A 3-dimensional Logical Expression

IF (Temperature is Low) AND (Temperature is Not Low) AND (Pressure is High)

THEN Speed is Very High

The IF part is a 3-dimensional logical expression such that

$A \; AND \; \bar{A} \; AND \; B$

where A is the truth value of (Temperature is Low), $\bar{A}$ is the truth value of (Temperature is Not Low) and B is the truth value of (Pressure is High).

(2) A 4-dimensional Logical Expression

IF (Temperature is Low) AND (Temperature is Not Low) AND (Pressure is High) AND (Pressure is Not High)

THEN Speed is High

The IF part is a 4-dimensional logical expression such that

$A \; AND \; \bar{A} \; AND \; B \; AND \; \bar{B}$

where A is the truth value of (Temperature is Low), $\bar{A}$ is the truth value of (Temperature is Not Low) , B is the truth value of (Pressure is High) and $\bar{B}$ is the truth value of (Pressure is Not High).

The question is "how to express CNF-DNF-based micro logical structures of m-dimensional logical expressions for $m \geq 3$?". Here, we show how to generate CNFs and DNFs for the above mentioned 3-dimensional Logical Expression and 4-dimensional Logical Expression.

At first, we use an AND_2^3 map to express $A \cap \bar{A} \cap B$ (see Table 2.15). Then we can easily get CNF and DNF expressions as follows,

$$CNF_2^3[a \triangle \bar{a} \triangle b] = \triangle_{i=1}^{64} C_i, \qquad (2.36)$$
$$DNF_2^3[a \triangle \bar{a} \triangle b] = (a \triangle b \triangle \bar{a}) \triangledown ((b \triangle a \triangle \bar{a}) \triangledown (\bar{a} \triangle b \triangle a) \triangledown$$
$$(b \triangle \bar{a} \triangle a) \triangledown ((\bar{a} \triangle a \triangle b) \triangledown ((a \triangle \bar{a} \triangle b), (2.37)$$

where $C_i \in OR_2^3$.

Table 2.15: AND_2^3 Map for $A \cap \bar{A} \cap B$.

	$\bar{a}$	a	b	b
$\bar{a} \triangle \bar{a}$	≤ 0.5	≤ 0.5	≤ 0.5	0.5
$\bar{a} \triangle a$	≤ 0.5	≤ 0.5	≤ 0.5	≤ 0.5
$a \triangle \bar{a}$	≤ 0.5	≤ 0.5	≤ 0.5	0.5
$a \triangle a$	≤ 0.5	≤ 0.5	≤ 0.5	≤ 0.5
$\bar{a} \triangle b$	≤ 0.5	≤ 0.5	≤ 0.5	≤ 0.5
$\bar{a} \triangle b$	≤ 0.5	0.5	≤ 0.5	≤ 0.5
$a \triangle b$	≤ 0.5	≤ 0.5	≤ 0.5	≤ 0.5
$a \triangle b$	0.5	≤ 0.5	≤ 0.5	≤ 0.5
$b \triangle \bar{a}$	≤ 0.5	≤ 0.5	≤ 0.5	≤ 0.5
$b \triangle a$	≤ 0.5	≤ 0.5	≤ 0.5	≤ 0.5
$b \triangle \bar{a}$	≤ 0.5	0.5	≤ 0.5	≤ 0.5
$b \triangle a$	0.5	≤ 0.5	≤ 0.5	≤ 0.5
$b \triangle b$	≤ 0.5	≤ 0.5	≤ 0.5	≤ 0.5
$b \triangle b$	≤ 0.5	≤ 0.5	≤ 0.5	≤ 0.5
$b \triangle b$	≤ 0.5	≤ 0.5	≤ 0.5	≤ 0.5
$b \triangle b$	≤ 0.5	≤ 0.5	≤ 0.5	≤ 0.5

Similarly, we can use an AND_2^4 map to express $A \cap \bar{A} \cap B \cap \bar{B}$. Then, we can get CNF and DNF expressions as follows,

$$CNF_2^4[a \triangle \bar{a} \triangle b \triangle \bar{b}] = \triangle_{i=1}^{256} C_i, \qquad (2.38)$$
$$DNF_2^4[a \triangle \bar{a} \triangle b \triangle \bar{b}] = (\bar{a} \triangle a \triangle \bar{b} \triangle b) \triangledown (\bar{a} \triangle a \triangle b \triangle \bar{b}) \triangledown$$
$$(\bar{a} \triangle \bar{b} \triangle a \triangle b) \triangledown (\bar{a} \triangle \bar{b} \triangle b \triangle a) \triangledown$$
$$(\bar{a} \triangle b \triangle a \triangle \bar{b}) \triangledown (\bar{a} \triangle b \triangle a \triangle \bar{b}) \triangledown$$
$$(a \triangle \bar{a} \triangle \bar{b} \triangle b) \triangledown (a \triangle \bar{a} \triangle b \triangle \bar{b}) \triangledown$$
$$(a \triangle \bar{b} \triangle \bar{a} \triangle b) \triangledown (a \triangle \bar{b} \triangle b \triangle \bar{a}) \triangledown$$
$$(a \triangle b \triangle \bar{a} \triangle \bar{b}) \triangledown (a \triangle b \triangle \bar{b} \triangle \bar{a}) \triangledown$$

$$(\bar{b} \triangle \bar{a} \triangle a \triangle b) \triangledown (\bar{b} \triangle \bar{a} \triangle b \triangle a) \triangledown$$
$$(\bar{b} \triangle a \triangle \bar{a} \triangle b) \triangledown (\bar{b} \triangle a \triangle b \triangle \bar{a}) \triangledown$$
$$(\bar{b} \triangle b \triangle \bar{a} \triangle a) \triangledown (\bar{b} \triangle b \triangle a \triangle \bar{a}) \triangledown$$
$$(b \triangle a \triangle \bar{a} \triangle \bar{b}) \triangledown (b \triangle a \triangle \bar{b} \triangle \bar{a}) \triangledown$$
$$(b \triangle \bar{b} \triangle \bar{a} \triangle a) \triangledown (b \triangle \bar{b} \triangle a \triangle \bar{a}), \quad (2.39)$$

where $C_i \in OR_2^4$.

2.5 Compensation of Universal Fuzzy CNF and Fuzzy DNF

According to the definitions in [50], for clarity we have the following definitions especially for 1-variable case.

Definition 1: Let a be a variable for $a \in [0,1]$, $\bar{a}$ be the complement of a such that $\bar{a} = 1 - a$. The 1-variable-1-dimensional fundamental phrase space is defined as
$AND_1^1 = \{a, \bar{a}\}$.
The 1-variable-1-dimensional fundamental phrase space is defined as
$AND_1^m = \{y_1 \cap y_2 \cap ... \cap y_m | y_1, y_2, ..., y_m \in AND_1^1, \text{ for } m \geq 2\}$.

Definition 2: Let a be a variable for $a \in [0,1]$, $\bar{a}$ be the complement of a such that $\bar{a} = 1 - a$. The 1-variable-1-dimensional fundamental clause space is defined as
$OR_1^1 = \{a, \bar{a}\}$.
The 1-variable-1-dimensional fundamental clause space is defined as
$OR_1^m = \{y_1 \cup y_2 \cup ... \cup y_m | y_1, y_2, ..., y_m \in OR_1^1, \text{ for } m \geq 2\}$.

Definition 3: A formula F is said to be in 1-variable-m-dimensional CNF if
$CNF_1^m(F) = C_1 \cap C_2 \cap ... \cap C_k$ for $k \geq 1$,
$CNF_1^m(1) = 1$,
and the fundamental clauses $C_i \in OR_1^m$ for $i = 1, 2, ..., k$ and $m \geq 1$.

Definition 4: A formula F is said to be in 1-variable-m-dimensional DNF if
$DNF_1^m(F) = P_1 \cup P_2 \cup ... \cup P_k$ for $k \geq 1$,
$DNF_1^m(0) = 0$,
and the fundamental phrases $P_i \in AND_1^m$ for $i = 1, 2, ..., k$ and $m \geq 1$.

2.5.1 Boolean Logic

The digital computer with binary strings of 1s and 0s stands as the emblem of the white and black and its triumph over the scientific mind. This faith in the black and the white, this *bivalence*, reaches back in the West to at least the ancient Greeks [61]. Aristotle's black-and-white laws of logic can be

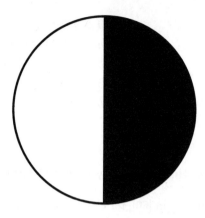

Figure 2.2. Aristotle's False(black)-True(white) Logic.

represented by the black-and-white figure (Figure 2.2). For example, we have two statements:
(1) It is a day (6 a.m.- 6 p.m.);
(2) It is not a day (6 p.m.- 6 a.m.).
If the truth value of the first statement is 1 (i.e., 100% true), then the truth value of the second statement must be 0 (i.e., 100% false). We use A and $\bar{A}$ to represent the truth values of the first affirmative statement and the second negative statement, respectively, then get the Boolean truth table for $A \cup \bar{A}$ (Table 2.16). Therefore, If $A = 1$ Then $\bar{A} = 0$, and If $A = 0$ Then $\bar{A} = 1$.

Table 2.16: 1-variable-1-dimensional Boolean Truth Table.

A	$A \cup A$	Phrases
0	1	$\bar{a}$
1	1	a

Suppose a crisp set's membership value $a \in \{0, 1\}$, $\bar{a} = 1 - a$, A and $\bar{A}$ are the degrees of truth of a and $\bar{a}$ respectively. $\cap$ and $\cup$ are the logical AND and OR for degrees of truth respectively, $\sqcap$ and $\sqcup$ are the computational AND and OR for degrees of membership respectively. According to the 1-variable-1-dimensional Boolean truth table (Table 2.16), we can easily get the 1-variable-1-dimensional CNF and DNF for $a \sqcup \bar{a}$ as follows,

$$CNF_1^1(a \sqcup \bar{a}) = 1, \qquad (2.40)$$
$$DNF_1^1(a \sqcup \bar{a}) = (a \sqcup \bar{a}). \qquad (2.41)$$

Sec. 2.5. Compensation of Universal Fuzzy CNF and Fuzzy DNF

Similarly, we can get

$$CNF_1^1(a \sqcap \bar{a}) = a \sqcap \bar{a}, \quad (2.42)$$
$$DNF_1^1(a \sqcap \bar{a}) = 0. \quad (2.43)$$

Since there are only two meaningful entries $\bar{a} \sqcap \bar{a}$ and $a \sqcap a$ whose truth values are all 1 in the 1-variable-2-dimensional Boolean Truth Table (Table 2.17), so we can get the 1-variable-2-dimensional CNF and DNF for $a \sqcup \bar{a}$ as follows,

$$CNF_1^2(a \sqcup \bar{a}) = 1, \quad (2.44)$$
$$DNF_1^2(a \sqcup \bar{a}) = (a \sqcap a) \sqcup (\bar{a} \sqcap \bar{a}). \quad (2.45)$$

Similarly, we can get

$$CNF_1^2(a \sqcap \bar{a}) = (a \sqcup a) \sqcap (\bar{a} \sqcup \bar{a}), \quad (2.46)$$
$$DNF_1^2(a \sqcap \bar{a}) = 0. \quad (2.47)$$

Table 2.17: 1-variable-2-dimensional Boolean Truth Table

A	A	$A \cup A$	Phrases
0	0	1	$\bar{a} \sqcap \bar{a}$
1	1	1	$a \sqcap a$

By induction, we can easily get the 1-variable-m-dimensional Boolean truth table (Table 2.18) which still has only two meaningful entries. According to the truth table (Table 2.18), we can get a 1-variable-m-dimensional CNFs and DNFs as follows,

$$CNF_1^m(a \sqcup \bar{a}) = 1, \quad (2.48)$$
$$DNF_1^m(a \sqcup \bar{a}) = \underbrace{(a \sqcap a...a \sqcap a)}_{m} \sqcup \underbrace{(\bar{a} \sqcap \bar{a}...\bar{a} \sqcap \bar{a})}_{m}. \quad (2.49)$$

Table 2.18: 1-variable-m-dimensional Boolean Truth Table.

A	A	...	A	A	$A \cup A$	Phrases
0	0	...	0	0	1	$\bar{a} \sqcap \bar{a}...\bar{a} \sqcap \bar{a}$
1	1	...	1	1	1	$a \sqcap a...a \sqcap a$

Similarly, we can get

$$CNF_1^m(a \sqcap \bar{a}) = \underbrace{(a \sqcup a...a \sqcup a)}_{m} \sqcap \underbrace{(\bar{a} \sqcup \bar{a}...\bar{a} \sqcup \bar{a})}_{m}, \quad (2.50)$$

$$DNF_1^m(a \sqcap \bar{a}) = 0. \quad (2.51)$$

It is easy to prove the following theorems,
Theorem 2.1: For $m \geq 1$, we have the identity laws,

$$CNF_1^m(a \sqcup \bar{a}) = DNF_1^m(a \sqcup \bar{a}) = 1, \quad (2.52)$$
$$CNF_1^m(a \sqcap \bar{a}) = DNF_1^m(a \sqcap \bar{a}) = 0. \quad (2.53)$$

Theorem 2.2: For $m \geq 1$, we have the conservation laws,

$$CNF_1^m(a \sqcup \bar{a}) + DNF_1^m(a \sqcap \bar{a}) = 1, \quad (2.54)$$
$$CNF_1^m(a \sqcup \bar{a}) + CNF_1^m(a \sqcap \bar{a}) = 1, \quad (2.55)$$
$$CNF_1^m(a \sqcap \bar{a}) + DNF_1^m(a \sqcup \bar{a}) = 1, \quad (2.56)$$
$$CNF_1^m(a \sqcap \bar{a}) + CNF_1^m(a \sqcup \bar{a}) = 1. \quad (2.57)$$

2.5.2 General Fuzzy Logic

Aristotle's black-white logical world looks perfect. But if a child asked Aristotle :"now is 6:00:00 a.m., is it a day or a night?", what would he say? The eastern philosophy, totally different from Aristotle's one, may solve this puzzle. Before 4602 B.C., Taichi (Figure 2.3) was created by ancient Chinese people [4]. Obviously, Taichi is totally different from Aristotle's black-white figure (Figure 2.2) in the logical sense. Taichi consists of Yin Fish (black) representing Yin and Yang Fish (white) representing Yang. Yin (Yang) Fish has a Yang (Yin) Eye represented by a small circle in Figure 2.3. The S-type curve implies (1) Yin can gradually be changed into Yang and Yang can gradually be changed into Yin; (2) the compensation of Yin and Yang; and (3) the nonlinear relation of Yin and Yang (see Figures 2.2 and 2.3). The principles of Taichi contain the principles of compensation of Yin and Yang. The Yang Eye in the Yin Fish means Yang is contained by Yin, and the Yin Eye in the Yang Fish means Yin is contained by Yang. In general, there exist a relative Yin containing Yang and a relative Yang containing Yin. Therefore, according to the principles of Taichi, we can say that if now is 6:00:00 a.m., it is a 50% day and a 50% night. As a result, Taichi is a perfect symbol of fuzzy logic.

Suppose a fuzzy set's membership value $a \in [0,1]$, $\bar{a} = 1 - a$, A and $\bar{A}$ are the degrees of truth of a and $\bar{a}$ respectively. $\sqcap$ and $\sqcup$ are the logical AND and

Sec. 2.5. Compensation of Universal Fuzzy CNF and Fuzzy DNF   27

Figure 2.3. Taichi Yin(black)-Yang(white) Logic.

OR for degrees of truth respectively, $\triangle$ and $\triangledown$ are the computational AND and OR for degrees of membership respectively. Since AND_n^m map completely describing all possible sub-concepts is logically equivalent to the corresponding n-variable-m-dimensional truth table [50], so both methods can get the same result. For clarity, we use both methods to generate CNFs and DNFs for $a \triangledown \bar{a}$ and $a \triangle \bar{a}$ as follows.

(1) 1-variable-1-dimensional CNFs and DNFs

According to the AND_1^1 map (Table 2.19) and the 1-variable-1-dimensional truth table (Table 2.20), we can easily get the 1-variable-1-dimensional CNF and DNF for $a \triangledown \bar{a}$ as follows,

$$CNF_1^1(a \triangledown \bar{a}) = 1, \qquad (2.58)$$
$$DNF_1^1(a \triangledown \bar{a}) = (a \triangledown \bar{a}). \qquad (2.59)$$

Table 2.19: 1-variable-1-dimensional AND_1^1 Map.

$\bar{a}$	a
≥ 0.5	≥ 0.5

Table 2.20: 1-variable-1-dimensional Fuzzy Truth Table.

A	$A \cup \bar{A}$	Phrases
$[0, 0.5]$	≥ 0.5	$\bar{a}$
$[0.5, 1]$	≥ 0.5	a

Similarly, we can get

$$CNF_1^1(a \triangle \bar{a}) = a \triangle \bar{a}, \qquad (2.60)$$
$$DNF_1^1(a \triangle \bar{a}) = 0. \qquad (2.61)$$

(2) 1-variable-2-dimensional CNFs and DNFs

Importantly, the 1-variable-2-dimensional Boolean truth table (Table 2.17) has only two entries since there exist excluded-middle law ($a \sqcup \bar{a} = 1$) and contradiction law ($a \sqcap \bar{a} = 0$), whereas the 1-variable-2-dimensional fuzzy truth table (Table 2.22) has 4 entries since $A \cap \bar{A} = 0.5$ and $\bar{A} \cap A = 0.5$ for $A = 0.5$. Therefore, the 50%-true-and-50%-false points are crucial for fuzzy logic. Since there are two phrases $\bar{a} \triangle a$ and $a \triangle \bar{a}$ whose truth values are all 0.5 in the AND_1^2 map (Table 2.21) and the 1-variable-2-dimensional truth table (Table 2.22), we have to use 50%-true-and-50%-false $\bar{a} \triangle a$ and $a \triangle \bar{a}$ to generate not only a DNF but also a CNF. Finally, we can get the 1-variable-2-dimensional CNF and DNF for $a \triangledown \bar{a}$ as follows,

$$CNF_1^2(a \triangledown \bar{a}) = (a \triangledown \bar{a}) \triangle (\bar{a} \triangledown a), \qquad (2.62)$$
$$DNF_1^2(a \triangledown \bar{a}) = (a \triangle \bar{a}) \triangledown (a \triangle a) \triangledown (\bar{a} \triangle \bar{a}) \triangledown (\bar{a} \triangle a). \qquad (2.63)$$

Table 2.21: 1-variable-2-dimensional AND_1^1 Map.

	$\bar{a}$	a
$\bar{a}$	≥ 0.5	0.5
a	0.5	≥ 0.5

Table 2.22: 1-variable-2-dimensional Fuzzy Truth Table.

A	$\bar{A}$	$A \cup \bar{A}$	Phrases
$[0, 0.5]$	$[0, 0.5]$	≥ 0.5	$\bar{a} \triangle \bar{a}$
$[0, 0.5]$	$[0.5, 1]$	0.5	$\bar{a} \triangle a$
$[0.5, 1]$	$[0, 0.5]$	0.5	$a \triangle \bar{a}$
$[0.5, 1]$	$[0.5, 1]$	≥ 0.5	$a \triangle a$

Similarly, we can get

$$CNF_1^2(a \triangle \bar{a}) = (a \triangledown \bar{a}) \triangle (a \triangledown a) \triangle (\bar{a} \triangledown \bar{a}) \triangle (\bar{a} \triangledown a), \qquad (2.64)$$
$$DNF_1^2(a \triangle \bar{a}) = (a \triangle \bar{a}) \triangledown (\bar{a} \triangle a). \qquad (2.65)$$

Sec. 2.5. Compensation of Universal Fuzzy CNF and Fuzzy DNF

(3) 1-variable-3-dimensional CNFs and DNFs
Similarly, we can get

$$CNF_1^3(a \triangledown \bar{a}) = (a \triangledown \bar{a} \triangledown \bar{a}) \triangle (a \triangledown \bar{a} \triangledown a) \triangle (a \triangledown a \triangledown \bar{a}) \triangle$$
$$(\bar{a} \triangledown \bar{a} \triangledown a) \triangle (\bar{a} \triangledown a \triangledown \bar{a}) \triangle (\bar{a} \triangledown a \triangledown a), \quad (2.66)$$

$$DNF_1^3(a \triangledown \bar{a}) = (a \triangle \bar{a} \triangle \bar{a}) \triangledown (a \triangle \bar{a} \triangle a) \triangledown (a \triangle a \triangle \bar{a}) \triangledown$$
$$(a \triangle a \triangle a) \triangledown (\bar{a} \triangle \bar{a} \triangle \bar{a}) \triangledown (\bar{a} \triangle \bar{a} \triangle a) \triangledown$$
$$(\bar{a} \triangle a \triangle \bar{a}) \triangledown (\bar{a} \triangle a \triangle a), \quad (2.67)$$

$$CNF_1^3(a \triangle \bar{a}) = (a \triangledown \bar{a} \triangledown \bar{a}) \triangle (a \triangledown \bar{a} \triangledown a) \triangle (a \triangledown a \triangledown \bar{a}) \triangle$$
$$(a \triangledown a \triangledown a) \triangle (\bar{a} \triangledown \bar{a} \triangledown \bar{a}) \triangle (\bar{a} \triangledown \bar{a} \triangledown a) \triangle$$
$$(\bar{a} \triangledown a \triangledown \bar{a}) \triangle (\bar{a} \triangledown a \triangledown a), \quad (2.68)$$

$$DNF_1^3(a \triangle \bar{a}) = (a \triangle \bar{a} \triangle \bar{a}) \triangledown (a \triangle \bar{a} \triangle a) \triangledown (a \triangle a \triangle \bar{a}) \triangledown$$
$$(\bar{a} \triangle \bar{a} \triangle a) \triangledown (\bar{a} \triangle a \triangle \bar{a}) \triangledown (\bar{a} \triangle a \triangle a). \quad (2.69)$$

(4) 1-variable-m-dimensional CNFs and DNFs
Therefore, for 1-variable-m-dimensional cases for $a \triangledown \bar{a}$, $\underbrace{\bar{a} \triangle \bar{a} ... \triangle \bar{a} \triangle \bar{a}}_{m}$ and $\underbrace{a \triangle a ... \triangle a \triangle a}_{m}$ have the truth value ≥ 0.5, but the other phrases have the same truth value 0.5. Therefore, 1-variable-m-dimensional DNF consists of all fundamental phrases in AND_1^m, and 1-variable-m-dimensional CNF consists of all fundamental clauses except $\underbrace{\bar{a} \triangledown \bar{a} ... \triangledown \bar{a} \triangledown \bar{a}}_{m}$ and $\underbrace{a \triangledown a ... \triangledown a \triangledown a}_{m}$ in OR_1^m.

2.5.3 m-dimensional Fuzzy CNFs and DNFs of $a \wedge \bar{a}$ and $a \vee \bar{a}$

By using $Min \wedge$ and $Max \vee$ to replace $\triangle$ and $\triangledown$ in the formulas (2.58)-(2.69), respectively, we get the following fuzzy CNFs and DNFs,

$$CNF_1^1(a \vee \bar{a}) = 1, \quad (2.70)$$
$$DNF_1^1(a \vee \bar{a}) = (a \vee \bar{a}), \quad (2.71)$$
$$CNF_1^1(a \wedge \bar{a}) = a \wedge \bar{a}, \quad (2.72)$$
$$DNF_1^1(a \wedge \bar{a}) = 0. \quad (2.73)$$
$$CNF_1^2(a \vee \bar{a}) = (a \vee \bar{a}) \wedge (\bar{a} \vee a), \quad (2.74)$$
$$DNF_1^2(a \vee \bar{a}) = (a \wedge \bar{a}) \vee (a \wedge a) \vee (\bar{a} \wedge \bar{a}) \vee (\bar{a} \wedge a), \quad (2.75)$$

$$CNF_1^2(a \wedge \bar{a}) = (a \vee \bar{a}) \wedge (a \vee a) \wedge (\bar{a} \vee \bar{a}) \wedge (\bar{a} \vee a), \qquad (2.76)$$

$$DNF_1^2(a \wedge \bar{a}) = (a \wedge \bar{a}) \vee (\bar{a} \wedge a). \qquad (2.77)$$

$$CNF_1^3(a \vee \bar{a}) = (a \vee \bar{a} \vee \bar{a}) \wedge (a \vee \bar{a} \vee a) \wedge (a \vee a \vee \bar{a}) \wedge$$
$$(\bar{a} \vee \bar{a} \vee a) \wedge (\bar{a} \vee a \vee \bar{a}) \wedge (\bar{a} \vee a \vee a), \qquad (2.78)$$

$$DNF_1^3(a \vee \bar{a}) = (a \wedge \bar{a} \wedge \bar{a}) \vee (a \wedge \bar{a} \wedge a) \vee (a \wedge a \wedge \bar{a}) \vee$$
$$(a \wedge a \wedge a) \vee (\bar{a} \wedge \bar{a} \wedge \bar{a}) \vee (\bar{a} \wedge \bar{a} \wedge a) \vee$$
$$(\bar{a} \wedge a \wedge \bar{a}) \vee (\bar{a} \wedge a \wedge a), \qquad (2.79)$$

$$CNF_1^3(a \wedge \bar{a}) = (a \vee \bar{a} \vee \bar{a}) \wedge (a \vee \bar{a} \vee a) \wedge (a \vee a \vee \bar{a}) \wedge$$
$$(a \vee a \vee a) \wedge (\bar{a} \vee \bar{a} \vee \bar{a}) \wedge (\bar{a} \vee \bar{a} \vee a) \wedge$$
$$(\bar{a} \vee a \vee \bar{a}) \wedge (\bar{a} \vee a \vee a), \qquad (2.80)$$

$$DNF_1^3(a \wedge \bar{a}) = (a \wedge \bar{a} \wedge \bar{a}) \vee (a \wedge \bar{a} \wedge a) \vee (a \wedge a \wedge \bar{a}) \vee$$
$$(\bar{a} \wedge \bar{a} \wedge a) \vee (\bar{a} \wedge a \wedge \bar{a}) \vee (\bar{a} \wedge a \wedge a). \qquad (2.81)$$

It is trivial to prove the following theorems,
Theorem 2.3: For $m = 1$, we have

$$CNF_1^1(a \vee \bar{a}) = 1, \qquad (2.82)$$

$$DNF_1^1(a \vee \bar{a}) = \begin{cases} 1-a & [0, 0.5] \\ a & [0.5, 1], \end{cases} \qquad (2.83)$$

$$CNF_1^1(a \wedge \bar{a}) = \begin{cases} a & [0, 0.5] \\ 1-a & [0.5, 1], \end{cases} \qquad (2.84)$$

$$DNF_1^1(a \wedge \bar{a}) = 0. \qquad (2.85)$$

For $m \geq 2$, we have the identity relations,

$$CNF_1^m(a \vee \bar{a}) = DNF_1^m(a \vee \bar{a}) = \begin{cases} 1-a & [0, 0.5] \\ a & [0.5, 1], \end{cases} \qquad (2.86)$$

$$CNF_1^m(a \wedge \bar{a}) = DNF_1^m(a \wedge \bar{a}) = \begin{cases} a & [0, 0.5] \\ 1-a & [0.5, 1]. \end{cases} \qquad (2.87)$$

Theorem 2.4: For $m = 1$, we have the conservation laws,

$$CNF_1^1(a \vee \bar{a}) + DNF_1^1(a \wedge \bar{a}) = 1, \qquad (2.88)$$

$$CNF_1^1(a \wedge \bar{a}) + DNF_1^1(a \vee \bar{a}) = 1. \qquad (2.89)$$

For $m \geq 2$, we have the conservation laws,

$$CNF_1^m(a \vee \bar{a}) + DNF_1^m(a \wedge \bar{a}) = 1, \qquad (2.90)$$

$$CNF_1^m(a \vee \bar{a}) + CNF_1^m(a \wedge \bar{a}) = 1, \qquad (2.91)$$
$$CNF_1^m(a \wedge \bar{a}) + DNF_1^m(a \vee \bar{a}) = 1, \qquad (2.92)$$
$$CNF_1^m(a \wedge \bar{a}) + CNF_1^m(a \vee \bar{a}) = 1. \qquad (2.93)$$

2.5.4 m-dimensional t-norm-t-conorm CNFs and DNFs

Now we discuss only two kinds of t-norms and t-conorms which are algebraic product & algebraic sum and bounded product & bounded sum [58,99].

Algebraic Product & Algebraic Sum

Algebraic product $\star$ of a and b for $a, b \in [0, 1]$ is defined by

$$a \star b = ab. \qquad (2.94)$$

Algebraic sum $\uplus$ of a and b for $a, b \in [0, 1]$ is defined by

$$a \uplus b = a + b - ab. \qquad (2.95)$$

By using algebraic product $\star$ and algebraic sum $\uplus$ to replace $\triangle$ and $\triangledown$ in the formulas (2.58)-(2.69), respectively, we have

$$DNF_1^1(a \uplus \bar{a}) = (a \uplus \bar{a}), \qquad (2.96)$$
$$CNF_1^1(a \uplus \bar{a}) = 1, \qquad (2.97)$$
$$DNF_1^1(a \star \bar{a}) = 0, \qquad (2.98)$$
$$CNF_1^1(a \star \bar{a}) = a \star \bar{a}. \qquad (2.99)$$
$$DNF_1^2(a \uplus \bar{a}) = (a \star \bar{a}) \uplus (a \star a) \uplus (\bar{a} \star \bar{a}) \uplus (\bar{a} \star a), \qquad (2.100)$$
$$CNF_1^2(a \uplus \bar{a}) = (a \uplus \bar{a}) \star (\bar{a} \uplus a), \qquad (2.101)$$
$$DNF_1^2(a \star \bar{a}) = (a \star \bar{a}) \uplus (\bar{a} \star a), \qquad (2.102)$$
$$CNF_1^2(a \star \bar{a}) = (a \uplus \bar{a}) \star (a \uplus a) \star (\bar{a} \uplus \bar{a}) \star (\bar{a} \uplus a). \qquad (2.103)$$
$$\begin{aligned}DNF_1^3(a \uplus \bar{a}) = & (a \star \bar{a} \star \bar{a}) \uplus (a \star \bar{a} \star a) \uplus (a \star a \star \bar{a}) \uplus \\ & (a \star a \star a) \uplus (\bar{a} \star \bar{a} \star \bar{a}) \uplus (\bar{a} \star \bar{a} \star a) \uplus \\ & (\bar{a} \star a \star \bar{a}) \uplus (\bar{a} \star a \star a),\end{aligned} \qquad (2.104)$$
$$\begin{aligned}CNF_1^3(a \uplus \bar{a}) = & (a \uplus \bar{a} \uplus \bar{a}) \star (a \uplus \bar{a} \uplus a) \star (a \uplus a \uplus \bar{a}) \star \\ & (\bar{a} \uplus \bar{a} \uplus a) \star (\bar{a} \uplus a \uplus \bar{a}) \star (\bar{a} \uplus a \uplus a),\end{aligned} \qquad (2.105)$$
$$\begin{aligned}DNF_1^3(a \star \bar{a}) = & (a \star \bar{a} \star \bar{a}) \uplus (a \star \bar{a} \star a) \uplus (a \star a \star \bar{a}) \uplus \\ & (\bar{a} \star \bar{a} \star a) \uplus (\bar{a} \star a \star \bar{a}) \uplus (\bar{a} \star a \star a),\end{aligned} \qquad (2.106)$$

$$CNF_1^3(a \star \bar{a}) = (a \uplus \bar{a} \uplus \bar{a}) \star (a \uplus \bar{a} \uplus a) \star (a \uplus a \uplus \bar{a}) \star$$
$$(a \uplus a \uplus a) \star (\bar{a} \uplus \bar{a} \uplus \bar{a}) \star (\bar{a} \uplus \bar{a} \uplus a) \star$$
$$(\bar{a} \uplus a \uplus \bar{a}) \star (\bar{a} \uplus a \uplus a). \qquad (2.107)$$

It is trivial to prove the following theorems,

Theorem 2.5:

$$DNF_1^1(a \uplus \bar{a}) = 1 - a(1-a), \qquad (2.108)$$
$$CNF_1^1(a \uplus \bar{a}) = 1, \qquad (2.109)$$
$$DNF_1^1(a \star \bar{a}) = 0. \qquad (2.110)$$
$$CNF_1^1(a \star \bar{a}) = a(1-a), \qquad (2.111)$$
$$DNF_1^2(a \uplus \bar{a}) = 1 - a(2-a)(1-a^2)(1-a+a^2)^2, \qquad (2.112)$$
$$CNF_1^2(a \uplus \bar{a}) = [1-a+a^2)]^2, \qquad (2.113)$$
$$DNF_1^2(a \star \bar{a}) = 1 - [1-a+a^2]^2. \qquad (2.114)$$
$$CNF_1^2(a \star \bar{a}) = a(2-a)(1-a^2)(1-a+a^2)^2, \qquad (2.115)$$
$$DNF_1^3(a \uplus \bar{a}) = 1 - (1-a^3)(3a-3a^2+a^3)(1-a^2+a^3)^3$$
$$(1-a+2a^2-a^3)^3, \qquad (2.116)$$
$$CNF_1^3(a \uplus \bar{a}) = (1-a^2+a^3)^3(1-a+2a^2-a^3)^3, \qquad (2.117)$$
$$DNF_1^3(a \star \bar{a}) = 1 - (1-a^2+a^3)^3(1-a+2a^2-a^3)^3, \qquad (2.118)$$
$$CNF_1^3(a \star \bar{a}) = (1-a^3)(3a-3a^2+a^3)(1-a^2+a^3)^3$$
$$(1-a+2a^2-a^3)^3. \qquad (2.119)$$

Theorem 2.6: For $1 \leq m \leq 3$, we have the conservation laws,

$$CNF_1^m(a \uplus \bar{a}) + DNF_1^m(a \star \bar{a}) = 1, \qquad (2.120)$$
$$CNF_1^m(a \star \bar{a}) + DNF_1^m(a \uplus \bar{a}) = 1. \qquad (2.121)$$

Bounded Product & Bounded Sum

Bounded product $\odot$ of a and b for $a, b \in [0, 1]$ is defined by

$$a \odot b = Max\{0, a+b-1\}. \qquad (2.122)$$

Bounded sum $\oplus$ of a and b for $a, b \in [0, 1]$ is defined by

$$a \oplus b = Min\{1, a+b\}. \qquad (2.123)$$

Sec. 2.5. Compensation of Universal Fuzzy CNF and Fuzzy DNF

By using bounded product $\odot$ and bounded sum $\oplus$ to replace $\triangle$ and $\triangledown$ in the formulas (2.58)-(2.69), respectively, we get the following bounded CNFs and DNFs,

$$CNF_1^1(a \oplus \bar{a}) = 1, \quad (2.124)$$
$$DNF_1^1(a \oplus \bar{a}) = (a \oplus \bar{a}), \quad (2.125)$$
$$CNF_1^1(a \odot \bar{a}) = a \odot \bar{a}, \quad (2.126)$$
$$DNF_1^1(a \odot \bar{a}) = 0. \quad (2.127)$$
$$CNF_1^2(a \oplus \bar{a}) = (a \oplus \bar{a}) \odot (\bar{a} \oplus a), \quad (2.128)$$
$$DNF_1^2(a \oplus \bar{a}) = (a \odot \bar{a}) \oplus (a \odot a) \vee (\bar{a} \odot \bar{a}) \oplus (\bar{a} \odot a), \quad (2.129)$$
$$CNF_1^2(a \odot \bar{a}) = (a \oplus \bar{a}) \odot (a \oplus a) \odot (\bar{a} \oplus \bar{a}) \odot (\bar{a} \oplus a), \quad (2.130)$$
$$DNF_1^2(a \odot \bar{a}) = (a \odot \bar{a}) \oplus (\bar{a} \odot a). \quad (2.131)$$
$$CNF_1^3(a \oplus \bar{a}) = (a \oplus \bar{a} \oplus \bar{a}) \odot (a \oplus \bar{a} \oplus a) \odot (a \oplus a \oplus \bar{a}) \odot$$
$$(\bar{a} \oplus \bar{a} \oplus a) \odot (\bar{a} \oplus a \oplus \bar{a}) \odot (\bar{a} \oplus a \oplus a), \quad (2.132)$$
$$DNF_1^3(a \oplus \bar{a}) = (a \odot \bar{a} \odot \bar{a}) \oplus (a \odot \bar{a} \odot a) \oplus (a \odot a \odot \bar{a}) \oplus$$
$$(a \odot a \odot a) \oplus (\bar{a} \odot \bar{a} \odot \bar{a}) \oplus (\bar{a} \odot \bar{a} \odot a) \oplus$$
$$(\bar{a} \odot a \odot \bar{a}) \oplus (\bar{a} \odot a \odot a), \quad (2.133)$$
$$CNF_1^3(a \odot \bar{a}) = (a \oplus \bar{a} \oplus \bar{a}) \odot (a \oplus \bar{a} \oplus a) \odot (a \oplus a \oplus \bar{a}) \odot$$
$$(a \oplus a \oplus a) \odot (\bar{a} \oplus \bar{a} \oplus \bar{a}) \odot (\bar{a} \oplus \bar{a} \oplus a) \odot$$
$$(\bar{a} \oplus a \oplus \bar{a}) \odot (\bar{a} \oplus a \oplus a), \quad (2.134)$$
$$DNF_1^3(a \odot \bar{a}) = (a \odot \bar{a} \odot \bar{a}) \oplus (a \odot \bar{a} \odot a) \oplus (a \odot a \odot \bar{a}) \oplus$$
$$(\bar{a} \odot \bar{a} \odot a) \oplus (\bar{a} \odot a \odot \bar{a}) \oplus (\bar{a} \odot a \odot a). \quad (2.135)$$

We can prove the following theorems,
Theorem 2.7:

$$CNF_1^1(a \oplus \bar{a}) = DNF_1^1(a \oplus \bar{a}) = 1, \quad (2.136)$$
$$CNF_1^1(a \odot \bar{a}) = DNF_1^1(a \odot \bar{a}) = 0. \quad (2.137)$$

For $m \geq 2$, we have

$$CNF_1^m(a \oplus \bar{a}) = 1, \quad (2.138)$$

$$DNF_1^m(a \oplus \bar{a}) = \begin{cases} 1 - ma & [0, \frac{1}{m}) \\ 0 & [\frac{1}{m}, \frac{m-1}{m}] \\ ma - m + 1 & (\frac{m-1}{m}, 1], \end{cases} \quad (2.139)$$

$$CNF_1^m(a \odot \bar{a}) = \begin{cases} ma & [0, \frac{1}{m}) \\ 1 & [\frac{1}{m}, \frac{m-1}{m}] \\ m - ma & (\frac{m-1}{m}, 1], \end{cases} \quad (2.140)$$

$$DNF_1^m(a \odot \bar{a}) = 0. \quad (2.141)$$

Proof:
(1) For $a \oplus \bar{a}$ case
(1.1) m=1

$$DNF_1^1(a \oplus \bar{a}) = a \oplus \bar{a} = Min\{1, a+1-a\} = 1 = CNF_1^1(a \oplus \bar{a}), (2.142)$$
$$CNF_1^1(a \odot \bar{a}) = a \odot \bar{a} = Max\{0, a+1-a-1\} = 0 = DNF_1^1(a \odot \bar{a})(2.143)$$

(1.2) $m \geq 2$

Since $CNF_1^m(a \oplus \bar{a})$ consists of all fundamental clauses except $\underbrace{\bar{a} \oplus \bar{a} ... \oplus \bar{a} \oplus \bar{a}}_{m}$ and $\underbrace{a \oplus a ... \oplus a \oplus a}_{m}$ in OR_1^m, so there is a sub-clause $a \oplus \bar{a}$ or a sub-clause $\bar{a} \oplus a$ in each clause of $CNF_1^m(a \oplus \bar{a})$. Because $a \oplus \bar{a} = \bar{a} \oplus a = 1$ and $1 \oplus b = b \oplus 1 = 1$ for $b \in \{a, \bar{a}\}$, consequently the value of each clause of $CNF_1^m(a \oplus \bar{a})$ is 1. As $1 \odot 1 = 1$, so $\underbrace{1 \odot 1 ... \odot 1 \odot 1}_{m} = 1$. Finally, we have

$$CNF_1^m(a \oplus \bar{a}) = 1. \quad (2.144)$$

Suppose $dnf_1^m(a \oplus \bar{a})$ has all fundamental phrases except $\underbrace{\bar{a} \odot \bar{a} ... \odot \bar{a} \odot \bar{a}}_{m}$ and $\underbrace{a \odot a ... \odot a \odot a}_{m}$ in AND_1^m, then we have

$$DNF_1^m(a \oplus \bar{a}) = dnf_1^m(a \oplus \bar{a}) \oplus (\underbrace{\bar{a} \odot \bar{a} ... \odot \bar{a} \odot \bar{a}}_{m}) \oplus (\underbrace{a \odot a ... \odot a \odot a}_{m}).$$

Similarly, we can get

$$dnf_1^m(a \oplus \bar{a}) = \underbrace{0 \oplus 0 ... \oplus 0 \oplus 0}_{m} = 0. \quad (2.145)$$

So we have

$$DNF_1^m(a \oplus \bar{a}) = (\underbrace{\bar{a} \odot \bar{a} ... \odot \bar{a} \odot \bar{a}}_{m}) \oplus (\underbrace{a \odot a ... \odot a \odot a}_{m}). \quad (2.146)$$

Sec. 2.5. Compensation of Universal Fuzzy CNF and Fuzzy DNF

Then we can prove the two following results,

$$\underbrace{\bar{a} \odot \bar{a} ... \odot \bar{a} \odot \bar{a}}_{m} = \begin{cases} 1 - ma & [0, \frac{1}{m}) \\ 0 & [\frac{1}{m}, 1], \end{cases} \quad (2.147)$$

$$\underbrace{a \odot a ... \odot a \odot a}_{m} = \begin{cases} 0 & [0, \frac{m-1}{m}] \\ ma - m + 1 & (\frac{m-1}{m}, 1]. \end{cases} \quad (2.148)$$

Finally, we have the conclusion

$$DNF_1^m(a \oplus \bar{a}) = \begin{cases} 1 - ma & [0, \frac{1}{m}) \\ 0 & [\frac{1}{m}, \frac{m-1}{m}] \\ ma - m + 1 & (\frac{m-1}{m}, 1]. \end{cases} \quad (2.149)$$

(2) For $a \odot \bar{a}$ case
Similarly, we can prove the following result,

$$CNF_1^m(a \odot \bar{a}) = \begin{cases} ma & [0, \frac{1}{m}) \\ 1 & [\frac{1}{m}, \frac{m-1}{m}] \\ m - ma & (\frac{m-1}{m}, 1], \end{cases} \quad (2.150)$$

$$DNF_1^m(a \odot \bar{a}) = 0. \quad (2.151)$$

Q.E.D.

Theorem 2.8: For $m = 1$, we have the conservation laws,

$$CNF_1^1(a \oplus \bar{a}) + DNF_1^1(a \odot \bar{a}) = 1, \quad (2.152)$$
$$CNF_1^1(a \oplus \bar{a}) + CNF_1^1(a \odot \bar{a}) = 1, \quad (2.153)$$
$$CNF_1^1(a \odot \bar{a}) + DNF_1^1(a \oplus \bar{a}) = 1, \quad (2.154)$$
$$CNF_1^1(a \odot \bar{a}) + CNF_1^1(a \oplus \bar{a}) = 1. \quad (2.155)$$

For $m \geq 2$, we have the conservation laws,

$$CNF_1^m(a \oplus \bar{a}) + DNF_1^m(a \odot \bar{a}) = 1, \quad (2.156)$$
$$CNF_1^m(a \odot \bar{a}) + DNF_1^m(a \oplus \bar{a}) = 1. \quad (2.157)$$

Proof: Trivial.

2.5.5 Relations in Fuzzy and t-norm-t-conorm CNFs and DNFs

In order to analyze the properties of various CNFs and DNFs based on different logics, we have calculated 1-variable-m-dimensional CNFs and DNFs. For clarity, we discuss them step by step.

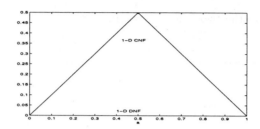

Figure 2.4. 1-variable-1-dimensional Fuzzy CNF and DNF for $a \wedge \bar{a}$.

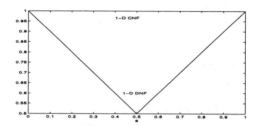

Figure 2.5. 1-variable-1-dimensional Fuzzy CNF and DNF for $a \vee \bar{a}$.

Fuzzy CNFs and DNFs

In this case, Zadeh's fuzzy logic is applied to CNFs and DNFs. The values of 1-variable-1-dimensional CNFs are greater than or equal to those of 1-variable-1-dimensional DNFs (see Figures 2.4 and 2.5). Interestingly, values of 1-variable-m-dimensional CNFs are equal to those of 1-variable-m-dimensional DNFs for $m \geq 2$ (see Figures 2.6 and 2.7).

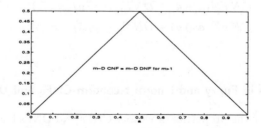

Figure 2.6. 1-variable-m-dimensional Fuzzy CNF and DNF for $a \wedge \bar{a}$.

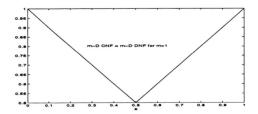

Figure 2.7. 1-variable-m-dimensional Fuzzy CNF and DNF for $a \vee \bar{a}$.

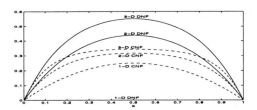

Figure 2.8. 1-variable-m-dimensional Algebraic CNF and DNF for $a \star \bar{a}$.

Algebraic CNFs and DNFs

In this case, the algebraic fuzzy logic is applied to CNFs and DNFs. Obviously, values of 1-variable-1-dimensional CNFs are greater than or equal to those of 1-variable-1-dimensional DNFs (see Figures 2.8 and 2.9). Additionally, values of 1-variable-m-dimensional CNFs are less than or equal to those of 1-variable-m-dimensional DNFs for $2 \leq m \leq 3$ (see Figures 2.8-2.10). Consequently, the divergences of 1-variable-m-dimensional CNFs and 1-variable-m-dimensional DNFs are positive for $m = 1$ and negative for $2 \leq m \leq 3$.

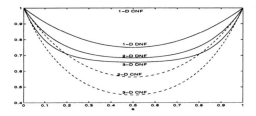

Figure 2.9. 1-variable-m-dimensional Algebraic CNF and DNF for $a \uplus \bar{a}$.

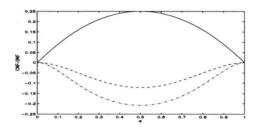

Figure 2.10. Divergence (CNF-DNF) of 1-variable-m-dimensional Algebraic CNF and DNF for $a \star \bar{a}$ or $a \uplus \bar{a}$ where solid line for $m = 1$, dashed line for $m = 2$ and dot-dashed line for $m = 3$.

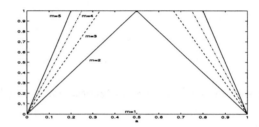

Figure 2.11. 1-variable-m-dimensional Bounded CNF and DNF for $a \odot \bar{a}$.

Bounded CNFs and DNFs

In this case, the bounded fuzzy logic is applied to CNFs and DNFs. The values of 1-variable-1-dimensional CNFs are equal to those of 1-variable-1-dimensional DNFs (see Figures 2.11 and 2.12). In addition, values of 1-variable-m-dimensional CNFs are greater than or equal to those of 1-variable-m-dimensional DNFs for $m \geq 2$ (see Figures 2.10 and 2.11). Interestingly, with the increasing of m for $m \geq 2$, the divergences of 1-variable-m-dimensional CNFs and 1-variable-m-dimensional DNFs will also increase (see Figures 2.11-2.13). Furthermore, when $m \to \infty$, the divergences of 1-variable-m-dimensional CNFs and 1-variable-m-dimensional DNFs will be 1 such that

$$\lim_{m \to \infty} [CNF_1^m(a \oplus \bar{a}) - DNF_1^m(a \oplus \bar{a})] = 1, \quad (2.158)$$

$$\lim_{m \to \infty} [CNF_1^m(a \odot \bar{a}) - DNF_1^m(a \odot \bar{a})] = 1. \quad (2.159)$$

Sec. 2.6. Summary

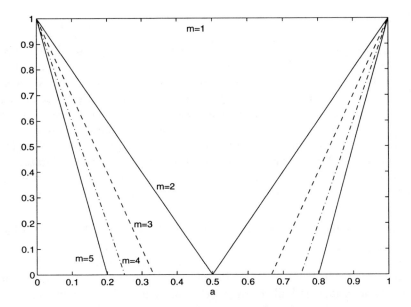

Figure 2.12. 1-variable-m-dimensional Bounded CNF and DNF for $a \oplus \bar{a}$.

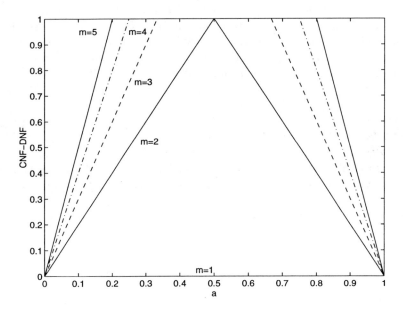

Figure 2.13. Divergence (CNF-DNF) of 1-variable-m-dimensional Bounded CNF and DNF for $a \odot \bar{a}$ or $a \oplus \bar{a}$.

2.6 Summary

Through analyzing the logical relations between Boolean truth tables and Karnaugh maps as well as Kaufmann's truth tables and Kaufmann's fuzzy maps, we have found that those truth tables and maps are no longer valid for general fuzzy logic such as T-norms and T-conorms because they lose many meaningful sub-concepts. Therefore, we propose degrees-of-truth-based universal fuzzy truth tables built on fundamental logical phrases and clauses [1] to avoid losing meaningful sub-concepts. Since universal fuzzy truth tables are logically equivalent to AND_n^m maps, we usually use AND_n^m maps to generate CNFs and DNFs for simplicity and convenience. Therefore, both universal fuzzy truth tables and AND_n^m maps are of conceptual completeness and logical consistency.

The 1-variable-m-dimensional CNFs and DNFs for $A \cup \bar{A}$ and $A \cap \bar{A}$ based on Boolean logic, fuzzy logic and t-norm-t-conorm logic are constructed by fundamental phrases and clauses in order to express micro logical structures of excluded middle law and contradiction law. From the philosophical point of view, Boolean logic is built on Aristotle's bivalent false-true logic, and fuzzy logic is based on Taichi Yin-Yang logic. So $A \cup \bar{A} = 1$ and $A \cap \bar{A} = 0$ are true for Boolean logic, but no longer true for fuzzy logic. From the logical point of view, a Boolean CNF is logically equivalent to a Boolean DNF, but a general fuzzy CNF is not logically equivalent to a general fuzzy DNF in most cases. From the methodological point of view, since AND_n^m map [1] and the universal fuzzy truth table are logically equivalent, so we can use one of them to generate CNFs and DNFs without losing any meaningful sub-concepts. The differences between 1-variable-m-dimensional CNFs and 1-variable-m-dimensional DNFs really indicate that people have various totally different kinds of thinking modes by using distinct fuzzy, vague or gray logics. For instance, Boolean-logic-based, fuzzy-logic-based and t-norm-t-conorm-logic-based CNFs and DNFs are totally different. Importantly, analysis can clearly indicate that whether excluded-middle law and contradiction Law hold is crucial for generating different kinds of CNFs and DNFs. Finally, how to effectively analyze the compensation between fuzzy DNF and fuzzy CNF and effectively apply it to fuzzy reasoning are important problems in fuzzy decision and other applications.

Chapter 3

Normal Fuzzy Reasoning Methodology

3.1 Primary Fuzzy Subsets

Because the conventional fuzzy sets theory takes all commonly used fuzzy sets as fundamental elements for fuzzy reasoning without splitting some complex fuzzy sets into more meaningful and more useful fuzzy subsets, the data granularity of conventionally used fuzzy sets is too low to contain heuristic information and mined knowledge (we will show examples in Chapter 5). Therefore, useful primary fuzzy sets are defined below in order to discover heuristic fuzzy knowledge in conventional fuzzy rule bases.

Definition 3.1: A monotonically non-increasing primary fuzzy set $\tilde{A}$ is defined by

$$\tilde{A} = \int_R \mu_{\tilde{A}}(x)/x, \qquad (3.1)$$

where $\mu_{\tilde{A}}(x)$ is a monotonically non-increasing function with respect to x for $x \in U$ (U is a universe of discourse).

Definition 3.2: A monotonically non-decreasing primary fuzzy set $\tilde{A}$ is defined by

$$\tilde{A} = \int_R \mu_{\tilde{A}}(x)/x, \qquad (3.2)$$

where $\mu_{\tilde{A}}(x)$ is a monotonically non-decreasing function with respect to x for $x \in U$ (U is a universe of discourse).

Definition 3.3: A constant primary fuzzy set $\tilde{A}$ is defined by

$$\tilde{A} = \int_R \mu_{\tilde{A}}(x)/x, \qquad (3.3)$$

where $\mu_{\tilde{A}}(x) = \alpha$ for a constant value $\alpha \in [0,1]$ and $x \in U$ (U is a universe of discourse).

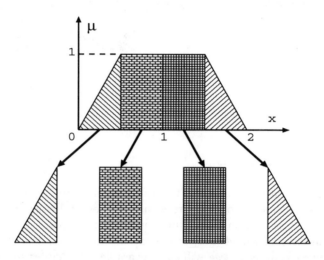

Figure 3.1. A trapezoidal fuzzy set with four primary fuzzy sets.

For example, the trapezoidal fuzzy set shown in Figure 3.1 consists of four primary fuzzy sets (the two crisp sets in Figure 3.1 are special cases of fuzzy sets). In general, we can use the four primary fuzzy sets in Figure 3.1 to construct various commonly used fuzzy sets shown in Figure 3.2. In Figure 3.2, three different kinds of triangular fuzzy sets and three different types of trapezoidal fuzzy sets are constructed by relevant primary fuzzy sets. Because the four fuzzy subsets are of different properties defined on different domains, it is necessary to take these primary fuzzy sets as different fundamental elements for fuzzy information processing.

3.2 The Variable-Input-Constant-Output (VICO) Problem

In order to analyze an unreasonable phenomenon called the VICO problem, let's look at the fuzzy air conditioner described in [50]. This fuzzy control system with 5 fuzzy rules can generate a crisp speed s for each crisp temperature t such that

Sec. 3.2. The Variable-Input-Constant-Output (VICO) Problem   43

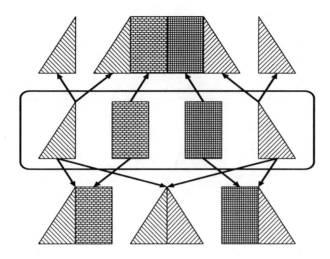

Figure 3.2. Commonly used fuzzy sets constructed by the primary fuzzy sets.

Rule: IF *Temperature* is *T*1 THEN *Speed* is *S*1
   IF *Temperature* is *T*2 THEN *Speed* is *S*2
   IF *Temperature* is *T*3 THEN *Speed* is *S*3
   IF *Temperature* is *T*4 THEN *Speed* is *S*4
   IF *Temperature* is *T*5 THEN *Speed* is *S*5
Input: *Temperature* is t
───────────────────────────────────────
Output :   *Speed* is s

where Ti and Si for $i = 1, 2, 3, 4, 5$ are fuzzy sets defined in Figure 3.3, t and s are crisp values.

For given $t = 50°F$ and $t = 60°F$, the conventional fuzzy reasoning method using the sup-min composition and the defuzzification scheme of center of gravity results in the non-intuitively constant speed of 30 rpm as shown in Figure 3.3. Heuristically, the air conditioner speed should increase when temperature increases according to the five fuzzy rules. Unreasonably, both the conventional method and the probabilistic approach generate constant speed (30 rpm) for any given temperature t for $50°F \leq t \leq 60°F$ [49]. This problem is called the variable-temperature-constant-speed problem, in general, the VICO problem.

Obviously, it is not effective to use conventional fuzzy sets as basic elements for fuzzy information processing.

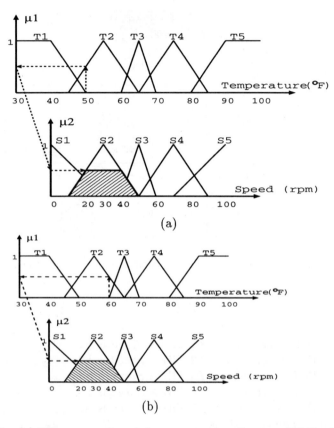

Figure 3.3. (a) The conventional fuzzy reasoning for $t = 50°F$. (b) The conventional fuzzy reasoning for $t = 60°F$.

3.3 Normal Fuzzy Reasoning (NFR)

For given m n-input-1-output fuzzy IF-THEN rules and n facts, what is a reasonable conclusion (i.e. an output fuzzy set B')? This problem is described below in detail:

IF x_1 is A_1^k and ... and x_n is A_n^k THEN y is B^k

x_1 is A_1' and ... and x_n is A_n'

$Conclusion:$ $\qquad\qquad\qquad\qquad$ y is B'

where trapezoidal-type fuzzy sets A_i^k for $i = 1, 2, ..., n$ and $k = 1, 2, ..., m$ are

Sec. 3.3. Normal Fuzzy Reasoning (NFR)

defined by

$$A_i^k = \int_{-\infty}^{+\infty} \mu_{A_i^k}(x_i)/x_i$$

$$= \int_{-\infty}^{a_i^k} \mu_{A_i^k}^l(x_i)/x_i \bigcup \int_{a_i^k-\alpha_i^k}^{a_i^k+\alpha_i^k} 1/x_i \bigcup \int_{a_i^k+\alpha_i^k}^{+\infty} \mu_{A_i^k}^r(x_i)/x_i, \quad (3.4)$$

in $U_i \subset R$, where $x_i \subset U_i$ are linguistic variables, and $\mu_{A_i^k}^l(x_i)$ and $\mu_{A_i^k}^r(x_i)$ are monotonically increasing functions and monotonically decreasing functions, respectively. Trapezoidal-type fuzzy sets B^k for $k = 1, 2, ..., m$ are defined by

$$B^k = \int_{-\infty}^{b^k-\beta^k} \mu_{B^k}^l(y)/y \bigcup \int_{b^k-\beta^k}^{b^k+\beta^k} 1/x \bigcup \int_{b^k+\beta^k}^{+\infty} \mu_{B^k}^r(y)/y, \quad (3.5)$$

in $V \subset R$, where $y \subset V$ is a linguistic variable, and $\mu_{B^k}^l(y)$ and $\mu_{B^k}^r(y)$ are monotonically increasing functions and monotonically decreasing functions, respectively. Trapezoidal-type fuzzy sets A_i' for $i = 1, 2, ..., n$ are defined by

$$A_i' = \int_{-\infty}^{+\infty} \mu_{A_i'}(x_i)/x_i$$

$$= \int_{-\infty}^{a_i'-\alpha_i'} \mu_{A_i'}^l(x_i)/x_i \bigcup \int_{a_i'-\alpha_i'}^{a_i'+\alpha_i'} 1/x_i \bigcup \int_{a_i'+\alpha_i'}^{+\infty} \mu_{A_i'}^r(x_i)/x_i, \quad (3.6)$$

where $\mu_{A_i'}^l(x_i)$ and $\mu_{A_i'}^r(x_i)$ are monotonically increasing functions and monotonically decreasing functions, respectively. Without loss of generality, suppose $\alpha_i^k > 0$, $\beta^k > 0$ and $\alpha_i' > 0$.

The normal fuzzy reasoning method is proposed as follows,
BEGIN
Step 1: Calculate the intersection of A_i^k and A_i' such that

$$\hat{A}_i^k = A_i^k \bigcap A_i', \quad (3.7)$$

$$\text{where } \mu_{\hat{A}_i^k}(x_i) = \min[\mu_{A_i^k}(x_i), \mu_{A_i'}(x_i)]. \quad (3.8)$$

Step 2: Calculate the compensatory strengths of firing rules. We have two typical methods as follows:

Method 1: Calculate Linear Compensatory Strengths

The linear compensatory strengths of firing rules are defined by

$$\lambda^k = \lambda_-(1-\gamma^k) + \lambda_+\gamma^k, \quad (3.9)$$

where γ^k are compensatory degrees, a pessimistic strength λ_- and an optimistic strength λ_+ are defined by

$$\lambda_- = min_{i=1}^n \{max[\mu_{\hat{A}_i^k}(x_i)]\}, \tag{3.10}$$

$$\lambda_+ = \frac{1}{n} \sum_{i=1}^n max[\mu_{\hat{A}_i^k}(x_i)]. \tag{3.11}$$

Method 2: Calculate Nonlinear Compensatory Strengths

The nonlinear compensatory strengths of firing rules are defined by

$$\lambda^k = (\lambda_-)^{1-\gamma^k}(\lambda_+)^{\gamma^k}, \tag{3.12}$$

where γ^k are the compensatory degrees, a pessimistic strength λ_- and an optimistic strength λ_+ are defined by

$$\lambda_- = \prod_{i=1}^n max[\mu_{\hat{A}_i^k}(x_i)], \tag{3.13}$$

$$\lambda_+ = \sqrt[n]{\prod_{i=1}^n max[\mu_{\hat{A}_i^k}(x_i)]}, \tag{3.14}$$

and positive membership functions such as Gaussian functions (i.e., $\mu_{A_i^k}(x_i) > 0$ and $\mu_{A_i'}(x_i) > 0$) are used in Step 1.

Therefore, we have

$$\lambda^k = [\prod_{i=1}^n max[\mu_{\hat{A}_i^k}(x_i)]]^{1-\gamma^k+\frac{\gamma^k}{n}}. \tag{3.15}$$

Step 3: Make the fuzzy equivalent mapping relations based on the trapezoidal-type membership functions:
CASE 1: If x_i increases, then y will increase.

$$\mu_{A_i^k}^l(x_i) = \mu_{B^k}^l(y), \quad for \ x_i \in (-\infty, a_i^k - \alpha_i^k), \tag{3.16}$$

$$\mu_{A_i^k}(x_i) = \mu_{B^k}(y) = 1, \quad for \ x_i \in [a_i^k - \alpha_i^k, a_i^k + \alpha_i^k], \tag{3.17}$$

$$\mu_{A_i^k}^r(x_i) = \mu_{B^k}^r(y), \quad for \ x_i \in (a_i^k + \alpha_i^k, +\infty). \tag{3.18}$$

After solving the above equations separately, we have:

$$x_i = \begin{cases} f_i^k(y) & for \ y \in (-\infty, b^k - \beta^k) \\ a_i^k + \frac{\alpha_i^k}{\beta^k}(y - b^k) & for \ y \in [b^k - \beta^k, b^k + \beta^k] \\ g_i^k(y) & for \ y \in (b^k + \beta^k, +\infty). \end{cases} \tag{3.19}$$

Sec. 3.3. Normal Fuzzy Reasoning (NFR)

CASE 2: If x_i increases, then y will decrease.

$$\mu^r_{A^k_i}(x_i) = \mu^l_{B^k}(y), \quad for \ x_i \in (a^k_i + \alpha^k_i, +\infty), \quad (3.20)$$

$$\mu_{A^k_i}(x_i) = \mu_{B^k}(y) = 1, \quad for \ x_i \in [a^k_i - \alpha^k_i, a^k_i + \alpha^k_i], \quad (3.21)$$

$$\mu^l_{A^k_i}(x_i) = \mu^r_{B^k}(y), \quad for \ x_i \in (-\infty, a^k_i - \alpha^k_i). \quad (3.22)$$

After solving the above equations separately, we have:

$$x_i = \begin{cases} v^k_i(y) & for \ y \in (-\infty, b^k - \beta^k) \\ a^k_i + \frac{\alpha^k_i}{\beta^k}(y + b^k) & for \ y \in [b^k - \beta^k, b^k + \beta^k] \\ w^k_i(y) & for \ y \in (b^k + \beta^k, +\infty). \end{cases} \quad (3.23)$$

CASE 3: If we don't really know the relations between x_i and y based on given data or knowledge, we may use two special parameters $wleft^k_i$ and $wright^k_i$ to generalize the above cases (we'll show an example at the end of this section).

Step 4: Map input $\hat{A}^k_i$ to corresponding output $\hat{B}^k_i$ such that
For Case 1:

$$\mu_{\hat{B}^k_i}(y) = \begin{cases} \mu_{\hat{A}^k_i}[f^k_i(y)] & for \ x_i \in (-\infty, a^k_i - \alpha^k_i) \\ \mu_{\hat{A}^k_i}[a^k_i + \frac{\alpha^k_i}{\beta^k}(y - b^k)] & for \ x_i \in [a^k_i - \alpha^k_i, a^k_i + \alpha^k_i] \\ \mu_{\hat{A}^k_i}[g^k_i(y)] & for \ x_i \in (a^k_i + \alpha^k_i, +\infty). \end{cases} \quad (3.24)$$

For Case 2:

$$\mu_{\hat{B}^k_i}(y) = \begin{cases} \mu_{\hat{A}^k_i}[v^k_i(y)] & for \ x_i \in (-\infty, a^k_i - \alpha^k_i) \\ \mu_{\hat{A}^k_i}[a^k_i + \frac{\alpha^k_i}{\beta^k}(y + b^k)] & for \ x_i \in [a^k_i - \alpha^k_i, a^k_i + \alpha^k_i] \\ \mu_{\hat{A}^k_i}[w^k_i(y)] & for \ x_i \in (a^k_i + \alpha^k_i, +\infty). \end{cases} \quad (3.25)$$

Step 5: Calculate the fuzzy cardinalities,

$$\eta^k_i = \int_V \mu_{\hat{B}^k_i}(y) dy, \quad (3.26)$$

the average fuzzy cardinalities are given by

$$\eta^k = \frac{1}{n} \sum_{i=1}^{n} \eta^k_i. \quad (3.27)$$

Step 6: Calculate the expected values of output fuzzy sets.

$$\delta^k_i = \frac{1}{\eta^k_i} \int_V \mu_{\hat{B}^k_i}(y) y \, dy. \quad (3.28)$$

The two typical methods to calculate the average expected values of output fuzzy sets:
(1) the weighted method:
$$\delta^k = \frac{\sum_{i=1}^n \delta_i^k \eta_i^k}{\sum_{i=1}^n \eta_i^k}, \qquad (3.29)$$

(2) the average method:
$$\delta^k = \frac{1}{n}\sum_{i=1}^n \delta_i^k. \qquad (3.30)$$

Step 7: The two typical methods to calculate the output expected value are given below,
(1) The fuzzy-cardinality-weighted method:
$$\delta = \frac{\sum_{k=1}^m \delta^k \eta^k \lambda^k}{\sum_{k=1}^m \eta^k \lambda^k}, \qquad (3.31)$$

(2) The firing-strength-weighted method:
$$\delta = \frac{\sum_{k=1}^m \delta^k \lambda^k}{\sum_{k=1}^m \lambda^k}. \qquad (3.32)$$

Step 8: The final conclusion is defined by a fuzzy set with the final fuzzy cardinality η such that
$$B' = \int_V \mu_{B'}(y)/y, \qquad (3.33)$$
$$\eta = \sum_{k=1}^m \eta^k. \qquad (3.34)$$

For an output Gaussian fuzzy set, we have
$$\mu_{B'}(y) = e^{-\frac{(y-\delta)^2}{2\sigma^2}}, \qquad (3.35)$$
$$\sigma = \frac{\eta}{\sqrt{2\pi}}. \qquad (3.36)$$

For an output triangular fuzzy set, we have
$$\mu_{B'}(y) = \begin{cases} 1 + \frac{y-\delta}{\eta} & \text{for } (\delta - \eta) \leq y \leq \delta \\ 1 - \frac{y-\delta}{\eta} & \text{for } \delta < y \leq (\delta + \eta) \\ 0 & \text{otherwise.} \end{cases} \qquad (3.37)$$

END.

3.4 Normal Fuzzy Controllers

In fuzzy control applications, there are four commonly used methods of fuzzy reasoning which are (1) Mamdani's method using the *minimum* fuzzy implication rule, (2) Larsen's method using the *product* fuzzy implication rule, (3) Tsukamoto's method with linguistic terms as monotonic membership functions and (4) Takagi-Sugeno's method with a fuzzy IF part and a crisp THEN part [73]. Since the frequently used Mamdani's method and Larsen's method may result in the VICO problem shown in Section 3.2, the new normal fuzzy reasoning method has been developed in the above section so as to overcome this weakness. Tsukamoto's method is not useful in many fuzzy control applications because it cannot use commonly used non-monotonic membership functions such as triangular functions and Gaussian functions. Takagi-Sugeno's method cannot extract commonly used fuzzy rules with both fuzzy inputs and fuzzy outputs from data, and also we cannot make a Takagi-Sugeno fuzzy system directly based on a fuzzy rule base. Our new fuzzy controller based on normal fuzzy reasoning (called normal fuzzy controller for simplicity) can solve the above mentioned problems.

A normal fuzzy control system with singleton fuzzifiers is described below,

$IF\ x_1\ is\ A_1^k\ and\ ...\ and\ x_n\ is\ A_n^k\ THEN\ y\ is\ B^k$

$\underline{x_1\ is\ a_1^*\ and\ ...\ and\ x_n\ is\ a_n^*}$

$Conclusion:\ \hspace{6em} y\ is\ b$

where a_i^* for $i = 1, 2, ..., n$ are crisp values, and b is a crisp value (note: A_i' and B' are fuzzy sets in Section 3.3, here crisp a_i^* and b are special cases of A_i' and B', respectively). A_i^k and B^k have been defined in Section 3.3.

Since singleton fuzzifiers are the simplest non-singleton fuzzifiers, the normal fuzzy reasoning method for non-singleton fuzzifiers in Section 3.3 can be used in fuzzy reasoning for singleton fuzzifiers. For clarity, the novel normal fuzzy reasoning method for singleton fuzzifiers is proposed as follows.
BEGIN
Step 1: Calculate the strengths of firing rules:

Method 1: Calculate Linear Compensatory Strengths

According to Eq. (2.2), we have

$$\lambda^k = \lambda_-(1-\gamma^k) + \lambda_+\gamma^k, \hspace{3em} (3.38)$$

where γ^k are compensatory degrees, a pessimistic strength λ_- and an optimistic strength λ_+ are defined by

$$\lambda_- = min_{i=1}^{n}[\mu_{A_i^k}(a_i^*)], \tag{3.39}$$

$$\lambda_+ = \frac{1}{n}\sum_{i=1}^{n}[\mu_{A_i^k}(a_i^*)]. \tag{3.40}$$

Method 2: Calculate Nonlinear Compensatory Strengths

According to Eq. (2.3), we have

$$\lambda^k = [\prod_{i=1}^{n}\mu_{A_i^k}(a_i^*)]^{1-\gamma^k+\frac{\gamma^k}{n}}, \tag{3.41}$$

where γ^k are the compensatory degrees.
Step 2: Get k outputs b_i^{*k} for given a_i^*:
CASE 1: If x_i increases, then y will increase.

$$\mu_{A_i^k}^l(a_i^*) = \mu_{B^k}^l(b_i^{*k}), \quad for \quad x_i \in (-\infty, a_i^k - \alpha_i^k), \tag{3.42}$$

$$b_i^{*k} = b^k + \frac{\beta^k}{\alpha_i^k}(a_i^* + a_i^k), \quad for \quad x_i \in [a_i^k - \alpha_i^k, a_i^k + \alpha_i^k], \tag{3.43}$$

$$\mu_{A_i^k}^r(a_i^*) = \mu_{B^k}^r(b_i^{*k}), \quad for \quad x_i \in (a_i^k + \alpha_i^k, +\infty). \tag{3.44}$$

CASE 2: If x_i increases, then y will decrease.

$$\mu_{A_i^k}^l(a_i^*) = \mu_{B^k}^r(b_i^{*k}), \quad for \quad x_i \in (-\infty, a_i^k - \alpha_i^k), \tag{3.45}$$

$$b_i^{*k} = b^k + \frac{\beta^k}{\alpha_i^k}(a_i^* - a_i^k), \quad for \quad x_i \in [a_i^k - \alpha_i^k, a_i^k + \alpha_i^k], \tag{3.46}$$

$$\mu_{A_i^k}^r(a_i^*) = \mu_{B^k}^l(b_i^{*k}), \quad for \quad x_i \in (a_i^k + \alpha_i^k, +\infty). \tag{3.47}$$

CASE 3: If we don't know the relations between x_i and y based on given data or knowledge, we may use two special parameters $wleft_i^k$ and $wright_i^k$ to generalize the above cases (we'll show an example at the end of this section).
Step 3: Calculate the average fuzzy cardinalities

$$\eta^k = \frac{1}{n}\sum_{i=1}^{n}\mu_{A_i^k}(a_i^*). \tag{3.48}$$

Step 4: Calculate the average expected values of output crisp values. The two typical methods are given below,

Sec. 3.4. Normal Fuzzy Controllers

(1) the weighted method:

$$\delta^k = \frac{\sum_{i=1}^{n} b_i^{*k} \mu_{A_i^k}(a_i^*)}{\sum_{i=1}^{n} \mu_{A_i^k}(a_i^*)}, \qquad (3.49)$$

(2) the average method:

$$\delta^k = \frac{1}{n} \sum_{i=1}^{n} b_i^{*k}. \qquad (3.50)$$

Step 5: The Normal Defuzzification Method (NDM) is used to calculate the output expected value b. The two typical NDMs are given below,
(1) The fuzzy-cardinality-weighted method:

$$b = \frac{\sum_{i=1}^{n} \delta^k \eta^k \lambda^k}{\sum_{i=1}^{n} \eta^k \lambda^k}, \qquad (3.51)$$

(2) The firing-strength-weighted method:

$$b = \frac{\sum_{k=1}^{m} \delta^k \lambda^k}{\sum_{k=1}^{m} \lambda^k}. \qquad (3.52)$$

END.

To understand the normal fuzzy reasoning, we use a simple example to show how to construct a 2-input-1-output normal fuzzy system if we don't know the relations between inputs and an output based on given data or knowledge (i.e., Case 3 in the above normal fuzzy reasoning method).

An $n \times n$ fuzzy rule base consists of $n \times n$ fuzzy rules such that

IF x_1 is A_{ij} and x_2 is B_{ij} THEN y is C_{ij}

x_1 is a and x_2 is b

$\overline{Conclusion: \qquad\qquad\qquad\qquad y \text{ is } c}$

where a, b and c are crisp values, input linguistic values A_{ij} and B_{ij} are defined as follows,

$$\mu_{A_{ij}}(x_1) = \begin{cases} 0 & \text{for } x_1 \leq (a_{ij} - \alpha_{ij}) \\ 1 + \frac{x_1}{\alpha_{ij}} - \frac{a_{ij}}{\alpha_{ij}} & \text{for } (a_{ij} - \alpha_{ij}) < x_1 \leq a_{ij} \\ 1 - \frac{x_1}{\alpha_{ij}} + \frac{a_{ij}}{\alpha_{ij}} & \text{for } a_{ij} < x_1 \leq (a_{ij} + \alpha_{ij}) \\ 0 & \text{for } x_1 > (a_{ij} + \alpha_{ij}), \end{cases} \qquad (3.53)$$

$$\mu_{B_{ij}}(x_2) = \begin{cases} 0 & \text{for } x_2 \leq (b_{ij} - \beta_{ij}) \\ 1 + \frac{x_2}{\beta_{ij}} - \frac{b_{ij}}{\beta_{ij}} & \text{for } (b_{ij} - \beta_{ij}) < x_2 \leq b_{ij} \\ 1 - \frac{x_2}{\beta_{ij}} + \frac{b_{ij}}{\beta_{ij}} & \text{for } b_{ij} < x_2 \leq (b_{ij} + \beta_{ij}) \\ 0 & \text{for } x_2 > (b_{ij} + \beta_{ij}), \end{cases} \quad (3.54)$$

output linguistic values C_{ij} are defined as follows,

$$\mu_{C_{ij}}(y) = \begin{cases} 0 & \text{for } y \leq (c_{ij} - \sigma_{ij}) \\ 1 + \frac{y}{\sigma_{ij}} - \frac{c_{ij}}{\sigma_{ij}} & \text{for } (c_{ij} - \sigma_{ij}) < y \leq c_{ij} \\ 1 - \frac{y}{\sigma_{ij}} + \frac{c_{ij}}{\sigma_{ij}} & \text{for } c_{ij} < y \leq (c_{ij} + \sigma_{ij}) \\ 0 & \text{for } y > (c_{ij} + \sigma_{ij}), \end{cases} \quad (3.55)$$

for $i, j = 1, 2, ..., n$.

In order to build relations between inputs and an output based on the given fuzzy rule base, heuristic parameters w_{ij}^a and w_{ij}^b are defined below,

$$w_{ij}^a = \begin{cases} w_{ij}^{aleft} & \text{for } x_1 \leq a_{ij} \\ w_{ij}^{aright} & \text{for } x_1 > a_{ij}, \end{cases} \quad (3.56)$$

$$w_{ij}^b = \begin{cases} w_{ij}^{bleft} & \text{for } x_2 \leq b_{ij} \\ w_{ij}^{bright} & \text{for } x_2 > b_{ij}. \end{cases} \quad (3.57)$$

According to the normal fuzzy reasoning method, a 2-input-1-output normal fuzzy system $f(x_1, x_2)$ based on the $n \times n$ fuzzy rules is given below,

$$f(x_1, x_2) = \frac{\sum_{i=1}^{n} \sum_{j=1}^{n} \lambda_{ij} [\mu_{A_{ij}}(x_1) \mu_{B_{ij}}(x_2)]^{1-\frac{\gamma_{ij}}{2}}}{\sum_{i=1}^{n} \sum_{j=1}^{n} [\mu_{A_{ij}}(x_1) \mu_{B_{ij}}(x_2)]^{1-\frac{\gamma_{ij}}{2}}}, \quad (3.58)$$

where

$$\lambda_{ij} = \{c_{ij} + \frac{\sigma_{ij}}{2} [\frac{w_{ij}^a(x_1 - a_{ij})}{\alpha_{ij}} + \frac{w_{ij}^b(x_2 - b_{ij})}{\beta_{ij}}]\}. \quad (3.59)$$

The heuristic parameters w_{ij}^a and w_{ij}^b can describe relations between inputs and an output.

For completeness, we discuss three cases:

Case 1: if x_i increases, then y will increase, we just choose $w_{ij}^a = 1$ and $w_{ij}^b = 1$.

Sec. 3.4. Normal Fuzzy Controllers

Case 2: if x_i increases, then y will decrease, we just choose $w_{ij}^a = -1$ and $w_{ij}^b = -1$.

Case 3: If we don't know the relations between x_i and y based on given data or knowledge, we have to develop a heuristic algorithm to find values of w_{ij}^{aleft}, w_{ij}^{aright}, w_{ij}^{bleft} and w_{ij}^{bright} based on the given fuzzy rule base. The heuristic algorithm is called the Knowledge Rediscovery Algorithm (KRA). Here, we use c_{ij} to represent output linguistic values C_{ij} for $i, j = 1, 2, ..., n$ in a fuzzy rule base. The KRA is given below.

Begin
  $i = 1$;
  for $(j = 1; j < (n+1); j++)$
    {if $(c_{ij} < c_{i+1j})$
      then $\{w_{ij}^{aleft} = 1;\ w_{ij}^{aright} = 1;\}$
      else {if $(c_{ij} > c_{i+1j})$
        then $w_{ij}^{aleft} = -1;\ w_{ij}^{aright} = -1;\}$
        else $w_{ij}^{aleft} = 0;\ w_{ij}^{aright} = 0;\}$
      }
    }
  for $(i = 2; i < (n-1); i++)$
    for $(j = 0; j < (n+1); j++)$
      {if $(c_{ij} < c_{i-1j})$
        then $\{w_{ij}^{aleft} = -1;\}$
        else {if $(c_{ij} > c_{i-1j})$
          then $\{w_{ij}^{aleft} = 1;\}$
          else $\{w_{ij}^{aleft} = 0;\}$
        }
      if $(c_{ij} < c_{i+1j})$
        then $\{w_{ij}^{aright} = 1;\}$
        else {if $(c_{ij} > c_{i+1j})$
          then $\{w_{ij}^{aright} = -1;\}$
          else $\{w_{ij}^{aright} = 0;\}$
        }
    }
  $i = n$;
  for $(j = 1; j < (n+1); j++)$
    {if $(c_{ij} < c_{i-1j})$
      then $\{w_{ij}^{aleft} = -1;\ w_{ij}^{aright} = -1;\}$

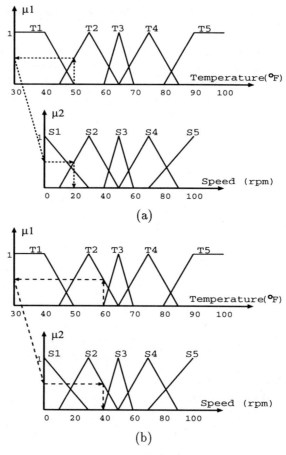

Figure 3.4. (a) The new normal fuzzy reasoning for $t = 50°F$. (b) The new normal fuzzy reasoning for $t = 60°F$.

$$\begin{aligned}
&\text{else } \{\text{if } (c_{ij} > c_{i-1j})\\
&\quad\text{then } \{w_{ij}^{aleft} = 1; w_{ij}^{aright} = 1;\}\\
&\quad\text{else } \{w_{ij}^{aleft} = 0; w_{ij}^{aright} = 0;\}\\
&\quad\}\\
&\}\\
&\text{End.}
\end{aligned}$$

Sec. 3.4. Normal Fuzzy Controllers   55

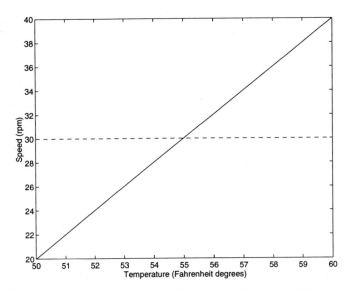

Figure 3.5. The speed generated by the NFR method (*solid line*) and that by both the conventional fuzzy reasoning method and the probabilistic approach (*dashed line*).

For two different input temperatures $t = 50°F$ and $t = 60°F$, the corresponding speeds of 20 rpm and 40 rpm (shown in Figures 3.4(a) and 3.4(b)) generated by the normal fuzzy reasoning method choosing $\gamma^k = 0$ and Eqs. (3.50) and (3.52) are more meaningful than constant 30 rpm (shown in Figures 3.3(a) and 3.3(b)) generated by both the conventional method and the probabilistic approach [49]. The reason is that we have discovered heuristic fuzzy knowledge which is "the higher temperature, the higher the speed" according to the five fuzzy rules. Our new method uses the rediscovered fuzzy knowledge to make heuristic fuzzy reasoning based on primary fuzzy subsets, but the conventional method doesn't because it makes fuzzy reasoning based on both non-primary fuzzy sets and primary ones. For completeness, the speeds generated by the normal fuzzy reasoning, the conventional Compositional Rule of Inference (CRI) and the probabilistic approach are shown in Figure 3.5. The main reason resulting in the constant speed of 30 rpm (see Figure 3.5) is the conventional fuzzy reasoning method treats the left side and right side of fuzzy sets $T2$ and $S2$ equally, so it doesn't take care of heuristic information such as "the higher temperature, the higher the speed".

Therefore, the conventional fuzzy reasoning method is not effective and many related defuzzification schemes have to be reconsidered because the conventional method takes non-primary fuzzy sets as fundamental elements for fuzzy information processing which results in losing useful knowledge and heuristic information. We'll show several examples in the following chapters. By splitting conventionally used non-primary fuzzy sets into more useful and heuristic primary fuzzy sets, the normal fuzzy reasoning methodology provides us with an efficient way to solve fuzzy reasoning problems such as the VICO problem. From the theoretical point of view, the novel methodology is a more reasonable foundation for fuzzy reasoning. The new normal fuzzy system is a generalized framework of many conventional fuzzy systems such as Takagi-Sugeno's fuzzy system [93,95], Wang's fuzzy system [103], Lin's fuzzy system [71,72] and the fuzzy system using Yager's level set method [26]. Interestingly, our normal fuzzy system can be constructed directly from given high-level fuzzy rules, but the conventional Takagi-Sugeno's fuzzy system has to be made from given low-level sample data by using a complex algorithm [77,79]. In this sense, the normal fuzzy reasoning methodology builds a bridge between Takagi-Sugeno's fuzzy system and other fuzzy systems. More importantly, the algorithm of normal fuzzy reasoning for fuzzy control is very simple because it doesn't need to calculate areas. Therefore, the normal fuzzy reasoning methodology, based on both logical and numerical methods, is useful and efficient in constructing robust fuzzy systems.

Chapter 4

Compensatory Genetic Fuzzy Neural Networks

After developing the normal fuzzy reasoning methodology, we begin now to apply it to a new hybrid system merging fuzzy logic, neural networks and genetic algorithms.

4.1 Introduction

An n-input-1-output normal fuzzy system has m fuzzy IF-THEN rules which are described by

$$\text{IF } x_1 \text{ is } A_1^k \text{ and } ... \text{ and } x_n \text{ is } A_n^k \text{ THEN } y \text{ is } B^k, \qquad (4.1)$$

where x_i and y are input and output fuzzy linguistic variables, respectively. Commonly used fuzzy linguistic values A_i^k and B^k are defined as follows (we may use triangular membership functions or others),

$$\mu_{A_i^k}(x_i) = exp[-(\frac{x_i - a_i^k}{\sigma_i^k})^2], \qquad (4.2)$$

$$\mu_{B^k}(y) = exp[-(\frac{y - b^k}{\eta^k})^2], \qquad (4.3)$$

where a_i^k and b^k are centers of membership functions of x_i and y, respectively, and σ_i^k and η^k are widths of membership functions of x_i and y, respectively, for $i = 1, 2, ..., n$ and $k = 1, 2, ..., m$.

Takagi-Sugeno's fuzzy system [95] is described below,

$$IF\ x_1\ is\ A_1^k\ and\ ...\ and\ x_n\ is\ A_n^k\ THEN\ y^k = p_0 + \sum_{i=1}^{n} p_i x_i, \qquad (4.4)$$

where p_i for $i = 0, 1, 2, ..., n$ are parameters.

Since the parameters in Takagi-Sugeno's fuzzy system have no physical meaning, Takagi-Sugeno's fuzzy system can not be constructed directly from given fuzzy rules with both a fuzzy IF part and a fuzzy THEN part. It has to be made from given low-level sample data by using a complex algorithm [93,95].

In order to overcome this weakness, according to the normal fuzzy reasoning we construct a compensatory normal fuzzy system $f(x_1, ..., x_n)$ directly based on given fuzzy rules as follows,

$$f(x_1, ..., x_n) = \frac{\sum_{k=1}^{m}(b^k + \frac{\eta^k}{n}\sum_{i=1}^{n}\frac{w_i^k(x_i - a_i^k)}{\sigma_i^k})[\prod_{i=1}^{n}\mu_{A_i^k}(x_i^k)]^{1-\gamma_k+\frac{\gamma_k}{n}}}{\sum_{k=1}^{m}[\prod_{i=1}^{n}\mu_{A_i^k}(x_i^k)]^{1-\gamma_k+\frac{\gamma_k}{n}}}, \qquad (4.5)$$

where heuristic parameters w_i^k are defined below,

$$w_i^k = \begin{cases} wleft_i^k & \text{for } x_i \le a_i^k \\ wright_i^k & \text{for } x_i > a_i^k, \end{cases} \qquad (4.6)$$

and the other parameters are defined in (4.2) and (4.3).

Interestingly, the output function $f(x_1, ..., x_n)$ also contains a linear combination of x_i for $i = 1, 2, ..., n$ like that in Takagi-Sugeno's fuzzy system since both input and output membership functions are all the same Gaussian functions. Especially, if input and output membership functions are different kinds of functions such as triangular and Gaussian functions, the output of a normal fuzzy system may have a nonlinear combination of x_i for $i = 1, 2, ..., n$. Therefore, a normal fuzzy system is a generalized framework of Takagi-Sugeno's fuzzy system. The normal fuzzy system $f(x_1, ..., x_n)$ will be used to construct a novel fuzzy neural network in the next section.

4.2 Fuzzy Neural Networks with Knowledge Discovery

In order to discover fuzzy rules (3.1) from data, we design a new fuzzy neural network called the Fuzzy Neural Network with Knowledge Discovery (FNNKD) (see Figure 4.1). The symbols and parameters related to (4.1) and (4.5) were

Sec. 4.2. Fuzzy Neural Networks with Knowledge Discovery

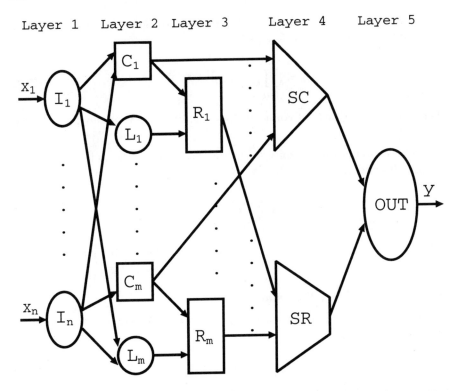

Figure 4.1. Architecture of a Fuzzy Neural Network with Knowledge Discovery.

already defined in Section 4.1. The FNNKD consists of 5 layers. The functions of fuzzy neurons in different layers are described layer by layer as follows:

Layer 1: Input Layer

Input neurons I_i in this layer are small oval nodes with simple mapping functions:

$$O_i = x_i, \qquad (4.7)$$

where O_i are outputs of input neurons I_i for $i = 1, 2, ..., n$.

Layer 2: Compensation and Linear Combination Layer

In this layer, there are two types of fuzzy neurons which are (1) compensatory neurons denoted by C_k and (2) linear combination neurons denoted by L_k for $k = 1, 2, ..., m$.

Compensatory neurons C_k in this layer are square nodes with compensatory mapping functions based on Eq. (3.41):

$$OC_k = [\prod_{i=1}^{n} \mu_{A_i^k}(x_i^k)]^{1-\gamma_k+\frac{\gamma_k}{n}}, \quad (4.8)$$

where OC_k are outputs of compensatory neurons C_k, γ_k stand for compensatory degrees.

Linear combination neurons L_k in this layer are circle nodes with linear mapping functions according to Eq. (3.48):

$$OL_k = b^k + \frac{\eta^k}{n} \sum_{i=1}^{n} \frac{w_i^k(x_i - a_i^k)}{\sigma_i^k}, \quad (4.9)$$

where OL_k are outputs of linear combination neurons L_k, the parameters are defined in (4.2) and (4.3).

Layer 3: Normal Fuzzy Reasoning Layer

Fuzzy rule neurons R_k in this layer are rectangle nodes with a product mapping functions:

$$O_k = OC_k OL_k, \quad (4.10)$$

where O_k are outputs of fuzzy rule neurons R_k.

Layer 4: Summation Layer

This layer consists of two neurons which are a compensatory summation neuron and a fuzzy rule summation neuron.

The compensatory summation neuron SC in this layer is a triangular node with a summation mapping function

$$O_{SC} = \sum_{k=1}^{m} OC_k, \quad (4.11)$$

where O_{SC} is an output of the compensatory summation neuron SC.

The fuzzy rule summation neuron SR in this layer is a trapezoidal node with a summation mapping function

$$O_{SR} = \sum_{k=1}^{m} O_k, \quad (4.12)$$

Sec. 4.3. Heuristic Genetic Learning Algorithm for a FNNKD

where O_{SR} is an output of the fuzzy rule summation neuron SR.

Layer 5: Output Layer

Finally, an output neuron OUT in this layer is a big oval node with a simple mapping function according to Eq. (3.51):

$$O_{OUT} = \frac{O_{SR}}{O_{SC}}, \qquad (4.13)$$

where O_{OUT} is an output of the output neuron OUT.

All parameters of a FNNKD include (1) a_i^k, (2) σ_i^k, (3) $wleft_i^k$, (4) $wright_i^k$, (5) b^k, (6) η^k and (7) γ^k for $i = 1, 2, ..., n$ and $k = 1, 2, ..., m$, therefore there are totally $m(4n+3)$ parameters. In order to learn these parameters based on sample data, we propose a hybrid learning algorithm in the next section.

4.3 Heuristic Genetic Learning Algorithm for a FNNKD

We have designed a Heuristic Genetic Learning Algorithm (HGLA) which is depicted in Figure 4.2.

The HGLA, the hybrid heuristic gradient descent learning algorithm, is described in detail below.

Given n-dimensional input data vectors x^p (i.e., $x^p = (x_1^p, x_2^p, ..., x_n^p)$) and 1-dimensional output data vector y^p for $p = 1, 2, ..., N$ and $N \geq m$ (m is the number of fuzzy rules). The normal fuzzy system $f(x_1^p, ..., x_n^p)$ was defined in Section 3.1. The objective function is defined by

$$E^p = \frac{1}{2}[f(x_1^p, ..., x_n^p) - y^p]^2. \qquad (4.14)$$

For simplicity, let E and f^p denote E^p and $f(x_1^p, ..., x_n^p)$, respectively.

Step 1: Begin.
Step 2:

If use only genetic algorithms to initialize parameters

Then goto Step 4;

Else goto Step 3.

Step 3: Heuristic-Knowledge-Based Initialization of Parameters.
Step 3.1: sort x_1^p for $p = 1, 2, ..., N$ in nondecreasing order, and change the order of corresponding $x_2^p, ..., x_n^p$ and y^p.
Step 3.2: partition sorted $x_1^p, x_2^p, ..., x_n^p$ and y^p for $p = 1, 2, ..., N$ into m data

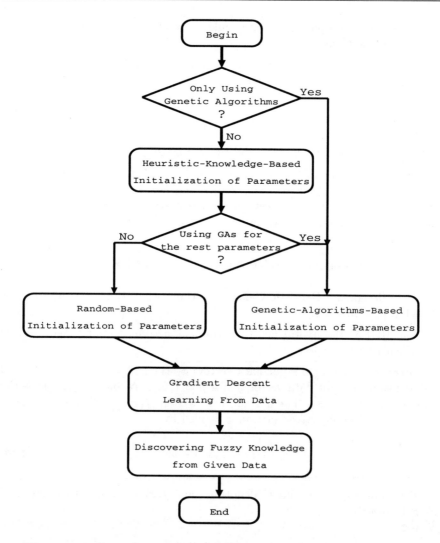

Figure 4.2. Procedure of the Heuristic Genetic Learning Algorithm.

segments (the kth segment data are used to initialize the parameters of the kth fuzzy rule), then calculate the heuristic initial values of parameters such that

Sec. 4.3. Heuristic Genetic Learning Algorithm for a FNNKD

$$a_i^k = \frac{1}{M} \sum_{j=K}^{I} x_i^j, \quad (4.15)$$

$$a_i^m = \frac{1}{L} \sum_{j=J}^{N} x_i^j, \quad (4.16)$$

$$b^k = \frac{1}{M} \sum_{j=K}^{I} y^j, \quad (4.17)$$

$$b^m = \frac{1}{L} \sum_{j=J}^{N} y^j, \quad (4.18)$$

$$\sigma_i^k = \alpha(Max_{j=K}^{I} x_i^j - Min_{j=K}^{I} x_i^j), \quad (4.19)$$

$$\sigma_i^m = \alpha(Max_{j=J}^{N} x_i^j - Min_{j=J}^{N} x_i^j), \quad (4.20)$$

$$\eta^k = \beta(Max_{j=K}^{I} y^j - Min_{j=K}^{I} y^j), \quad (4.21)$$

$$\eta^m = \beta(Max_{j=J}^{N} y^j - Min_{j=J}^{N} y^j), \quad (4.22)$$

where $i = 1, 2, ..., n$, $k = 1, 2, ..., m-1$, $M = \frac{N}{m}$, $L = N \mod m$, $K = 1 + (k-1)M$, $I = kM$ and $J = 1 + (m-1)M$.

Step 3.3:

If $\sigma_i^k = 0$, then

$$\sigma_i^k = \frac{\alpha}{m}(Max_{j=1}^{N} x_i^j - Min_{j=1}^{N} x_i^j), \quad (4.23)$$

If $\eta^k = 0$, then

$$\eta^k = \frac{\beta}{m}(Max_{j=1}^{N} y^j - Min_{j=1}^{N} y^j), \quad (4.24)$$

where $i = 1, 2, ..., n$, $k = 1, 2, ..., m$.

Step 3.4:

If $\sigma_i^k = 0$, then

$$\sigma_i^k = \frac{x_i^1}{m}, \quad (4.25)$$

If $\eta^k = 0$, then

$$\eta^k = \frac{y^1}{m}, \quad (4.26)$$

where $i = 1, 2, ..., n$, $k = 1, 2, ..., m$.

Step 3.5:

Set $\gamma^k = \gamma_0$ for $0 \leq \gamma_0 \leq 1$ such as $\gamma_0 = 0.5$, for $k = 1, 2, ..., m$.

Step 3.6:

If randomly set $wleft_i^k$ and $wright_i^k$

Then set $wleft_i^k$ and $wright_i^k$ with random numbers for $i = 1, 2, ..., n$, $k = 1, 2, ..., m$. Goto Step 4;

Else goto Step 4.

Step 4: Genetic-Algorithms–Based Initialization of Parameters.

Step 4.1:

If we use genetic algorithms to adjust only $wleft_i^k$ and $wright_i^k$

Then initialize a population of chromosomes for $wleft_i^k$ and $wright_i^k$ represented by strings of binary digits (1's and 0's).

Else initialize a population of chromosomes which encode the uninitialized parameters of a FNNKD by strings of binary digits (1's and 0's).

Step 4.2: Calculate the fitness function.

Step 4.3: Select better parent chromosomes in the population with higher fitness values according to the mechanics of natural selection and evolution.

Step 4.4: Perform the crossover and mutation operations to generate a new generation (users can define the number of crossover points and a mutation rate).

Step 4.5:

If the stopping criterion (such as if the iteration number is larger than the given number of generations) is met

Then goto Step 5;

Else goto Step 4.2.

Step 5: Gradient Descent Learning From Data.

For simplicity, we define γ, ϕ and θ as follows,

$$\gamma = 1 - \gamma_k + \frac{\gamma_k}{n}, \qquad (4.27)$$

$$\phi = \frac{1}{n}\sum_{i=1}^{n} \frac{w_i^k(x_i - a_i^k)}{\sigma_i^k}, \qquad (4.28)$$

Sec. 4.3. Heuristic Genetic Learning Algorithm for a FNNKD

$$\theta = \frac{(f^p - y^p)z_k^\gamma}{\sum_{k=1}^m z_k^\gamma}, \qquad (4.29)$$

where

$$z_k = \prod_{i=1}^n \mu_{A_i^k}(x_i^k), \qquad (4.30)$$

and the w_i^k are defined by (3.6). Then we can get the following learning algorithm for $i = 1, 2, ..., n$, $k = 1, 2, ..., m$, $p = 1, 2, ..., N$, $t = 0, 1, 2, ...$ and a learning rate $\lambda > 0$.

Step 5.1: Train b^k

$$\frac{\partial E}{\partial b^k}\bigg|_t = \theta\bigg|_t, \qquad (4.31)$$

$$b^k(t+1) = b^k(t) - \lambda \frac{\partial E}{\partial b^k}\bigg|_t. \qquad (4.32)$$

Step 5.2: Train η^k

$$\frac{\partial E}{\partial \eta^k}\bigg|_t = \theta\phi, \qquad (4.33)$$

$$\eta^k(t+1) = \eta^k(t) - \lambda \frac{\partial E}{\partial \eta^k}\bigg|_t. \qquad (4.34)$$

Step 5.3: Train a_i^k

$$\frac{\partial E}{\partial a_i^k}\bigg|_t = \frac{\theta}{n(\sigma_i^k)^2}[2n\gamma(x_i^p - a_i^k)(b^k - f^p + \eta^k\phi) - \eta^k \sigma_i^k w_i^k], \quad (4.35)$$

$$a_i^k(t+1) = a_i^k(t) - \lambda \frac{\partial E}{\partial a_i^k}\bigg|_t. \qquad (4.36)$$

Step 5.4: Train σ_i^k

$$\frac{\partial E}{\partial \sigma_i^k}\bigg|_t = \frac{\partial E}{\partial a_i^k} \frac{(x_i^p - a_i^k)}{\sigma_i^k}, \qquad (4.37)$$

$$\sigma_i^k(t+1) = \sigma_i^k(t) - \lambda \frac{\partial E}{\partial \sigma_i^k}\bigg|_t. \qquad (4.38)$$

Step 5.5 : Train $wleft_i^k$ or $wright_i^k$
IF $x_i^p \leq a_i^k$

THEN

$$\frac{\partial E}{\partial wleft_i^k}\bigg|_t = \frac{\theta \eta^k (x_i^p - a_i^k)}{n\sigma_i^k}, \qquad (4.39)$$

$$wleft_i^k(t+1) = wleft_i^k(t) - \lambda \frac{\partial E}{\partial wleft_i^k}\bigg|_t . \qquad (4.40)$$

ELSE

$$\frac{\partial E}{\partial wright_i^k}\bigg|_t = \frac{\theta \eta^k (x_i^p - a_i^k)}{n\sigma_i^k}, \qquad (4.41)$$

$$wright_i^k(t+1) = wright_i^k(t) - \lambda \frac{\partial E}{\partial wright_i^k}\bigg|_t . \qquad (4.42)$$

Step 5.6: Train γ^k

At first, we define

$$\gamma^k = \frac{c_k^2}{c_k^2 + d_k^2}, \qquad (4.43)$$

then we have

$$\frac{\partial E}{\partial c_k}\bigg|_t = \frac{2\theta(1-n)c_k d_k^2 (b^k - f^p + \eta^k \phi) ln(z_k)}{n(c_k^2 + d_k^2)}, \qquad (4.44)$$

$$c_k(t+1) = c_k(t) - \lambda \frac{\partial E}{\partial c_k}\bigg|_t, \qquad (4.45)$$

$$\frac{\partial E}{\partial d_k}\bigg|_t = \frac{2\theta(n-1)c_k^2 d_k (b^k - f^p + \eta^k \phi) ln(z_k)}{n(c_k^2 + d_k^2)}, \qquad (4.46)$$

$$d_k(t+1) = d_k(t) - \lambda \frac{\partial E}{\partial d_k}\bigg|_t, \qquad (4.47)$$

$$\gamma^k(t+1) = \frac{[c_k(t+1)]^2}{[c_k(t+1)]^2 + [d_k(t+1)]^2}. \qquad (4.48)$$

Step 6: Discovering Fuzzy Knowledge (Fuzzy IF-THEN Rules) from Data.

Now all parameters for a FNNKD have been adjusted, i.e., all m fuzzy rules described by (4.1) have been discovered from training data. Finally, the trained FNNKD can generate new values for new given input data.

Step 7: End.

4.4 Feature Expressions of Trapezoidal-type Fuzzy Sets

Since triangular and Gaussian fuzzy sets are very useful kinds of fuzzy sets, we now analyze how to express them by crisp features which can effectively be used in a fuzzy neural network. For completeness, we have 4 definitions below.

Definition 4.1: A trapezoidal quaternion (a,b,c,d) is a geometric expression of a trapezoidal fuzzy set $\tilde{A}$ which is defined by

$$\mu_{\tilde{A}}(x) = \begin{cases} \frac{x-c}{a-b-c} & \text{for } c \leq x < (a-b) \\ 1 & \text{for } (a-b) \leq x \leq (a+b) \\ \frac{d-x}{d-a-b} & \text{for } (a+b) < x \leq d \\ 0 & \text{otherwise.} \end{cases} \quad (4.49)$$

Definition 4.2: A Gaussian quaternion $(\alpha, \beta, \delta_-, \delta_+)$ is a geometric expression of a trapezoidal-Gaussian fuzzy set $\tilde{B}$ which defined by

$$\mu_{\tilde{B}}(x) = \begin{cases} e^{-\frac{(x-\alpha+\beta)^2}{\delta_-^2}} & \text{for } x < (\alpha-\beta) \\ 1 & \text{for } (\alpha-\beta) \leq x \leq (\alpha+\beta) \\ e^{-\frac{(x-\alpha-\beta)^2}{\delta_+^2}} & \text{for } x > (\alpha+\beta). \end{cases} \quad (4.50)$$

Definition 4.3: The fuzzy information capacity of a fuzzy set $\tilde{A} = \int_R \mu_{\tilde{A}}(x)/x$ in the interval of $[x_1, x_2]$ is defined by

$$\rho = \int_{x_1}^{x_2} \mu_{\tilde{A}}(x)dx. \quad (4.51)$$

Definition 4.4: A trapezoidal fuzzy information capacity quaternion denoted by $[a, b, \rho^-, \rho^+]$ is a characteristic expression of a trapezoidal-type fuzzy set $\tilde{A}$ which is defined by

$$\mu_{\tilde{A}}(x) = \begin{cases} f^-(x) & \text{for } x < (a-b) \\ 1 & \text{for } (a-b) \leq x \leq (a+b) \\ f^+(x) & \text{for } x > (a+b), \end{cases} \quad (4.52)$$

where

$$\rho^- = \int_{-\infty}^{a-b} f^-(x)dx, \quad (4.53)$$

$$\rho^+ = \int_{a+b}^{+\infty} f^+(x)dx, \quad (4.54)$$

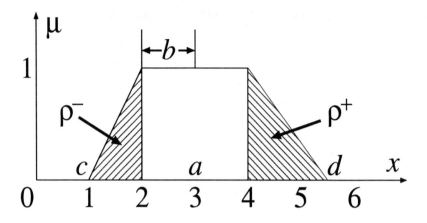

Figure 4.3. Fuzzy features of a trapezoidal fuzzy set.

where $f^-(x)$ and $f^+(x)$ are a monotonically nondecreasing function with respect to x and a monotonically non-increasing function with respect to x, respectively, $0 \leq f^-(x) \leq 1$ and $0 \leq f^+(x) \leq 1$.

For example, a trapezoidal fuzzy set is defined in Figure 4.3. Obviously, a trapezoidal quaternion (a,b,c,d) of the trapezoidal fuzzy set is (3,1,1,5.5), and a trapezoidal fuzzy information capacity quaternion $[a, b, \rho^-, \rho^+]$ is [3,1,0.5,0.75].

In general, a complex trapezoidal-type fuzzy set can be typically represented by crisp features a and b which represent a hard core of the fuzzy set and fuzzy features ρ^- and ρ^+ which represent soft left and right tails of the fuzzy set, respectively.

4.5 Crisp-Fuzzy Neural Networks (CFNN)

A 2-fuzzy input-1 fuzzy output CFNN (see Figure 4.4) consists of 3 layers. The functions of different layers are described layer by layer as follows:

Layer 1: Crisp and Fuzzy Features Generation Layer

In this layer, trapezoidal-type fuzzy sets X_1 and X_2 are transformed to corresponding fuzzy information capacity quaternions. In a fuzzy information capacity quaternion $[a, b, \rho^-, \rho^+]$, a and b represent crisp information contained in the input trapezoidal-type fuzzy set, and ρ^- and ρ^+ represent fuzzy information contained in the input trapezoidal-type fuzzy set. As a result, an input

Sec. 4.5. Crisp-Fuzzy Neural Networks (CFNN)

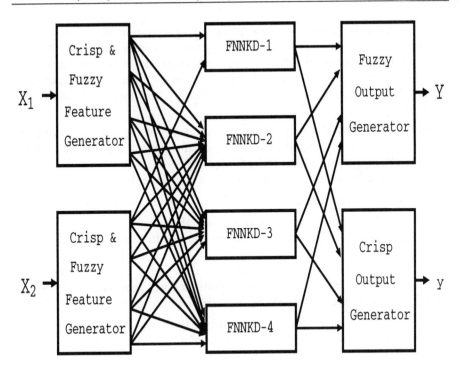

Figure 4.4. Architecture of a Crisp-Fuzzy Neural Network.

trapezoidal-type fuzzy set is represented by crisp features a and b which describe a *crisp* core of the fuzzy set and fuzzy features ρ^- and ρ^+ which describe *fuzzy* left and right tails of the fuzzy set, respectively. Importantly, an input fuzzy set has been represented by 4 crisp values which can easily be processed by a fuzzy or crisp neural network. Finally, the output of the first layer is a fuzzy information capacity quaternion $[a, b, \rho^-, \rho^+]$.

Layer 2: Neuro-fuzzy Reasoning with Crisp-Inputs and Crisp-Fuzzy-Inputs Layer

This layer consists of 4 dedicated FNNKDs (i.e., FNNKD-1, FNNKD-2, FNNKD-3 and FNNKD-4). FNNKD-1 is a 2-input-1-output fuzzy neural network which generates a crisp center α of an output fuzzy set based on 2 input centers of input fuzzy sets X_1 and X_2 by using k_1 fuzzy rules. FNNKD-2 is an 8-input-1-output fuzzy neural network which generates a crisp width β of an output fuzzy set based on 8 features of input fuzzy sets X_1 and X_2 by using k_2 fuzzy rules. FNNKD-3 is an 8-input-1-output fuzzy neural network which generates a left fuzzy information capacity ρ_y^- of an output fuzzy set based on 8 features of input fuzzy sets X_1 and X_2 by using k_3 fuzzy rules. FNNKD-4 is an 8-input-1-output fuzzy neural network which generates a right fuzzy information capacity ρ_y^+ of an output fuzzy set based on 8 features of input fuzzy sets X_1 and X_2 by using k_4 fuzzy rules. Finally, the output of this layer is $[\alpha, \beta, \rho_y^-, \rho_y^+]$.

In layer 2, FNNKD-1 plays the most important role in making better fuzzy reasoning because it generates a center of an output fuzzy set. FNNKD-1 and FNNKD-2 generate hard (crisp) information of fuzzy reasoning, whereas FNNKD-3 and FNNKD-4 generate soft (fuzzy) information of fuzzy reasoning.

Layer 3: Defuzzification Output and/or Fuzzy Output Layer

Case 1: If we only want a fuzzy output, the fuzzy output generator can easily transform $[\alpha, \beta, \rho_y^-, \rho_y^+]$ to a simple trapezoidal fuzzy set or a complex trapezoidal-type fuzzy set depending on a user's need. Actually, the output is nothing but a fuzzy reasoning conclusion represented by a trapezoidal-type fuzzy set which is much simpler than multiple fuzzy conclusions represented by multiple fuzzy sets generated by a conventional fuzzy reasoning method.

Case 2: If we only want a crisp output, the crisp output generator can perform simple defuzzification to generate a crisp value based on a trapezoidal-type fuzzy set represented by $[\alpha, \beta, \rho_y^-, \rho_y^+]$.

Obviously, a CFNN is a general fuzzy neural network which is able to process crisp or fuzzy input data and generate crisp or fuzzy output data. Therefore, one CFNN has all the functions of FIFO, FICO, CIFO and CICO fuzzy neural networks. In summary, a CFNN has a lot of abilities which are (1) learning fuzzy rules from either crisp sample data or fuzzy sample data; (2) compressing a fuzzy rule base with a large number of fuzzy rules to one with a small number of fuzzy rules (see Chapter 7); (3) expanding a sparse fuzzy rule base to a non-sparse fuzzy rule base (see Chapter 7); (4) making fuzzy reasoning for either crisp data or fuzzy data; (5) doing simple defuzzification because every fuzzy conclusion is nothing but a trapezoidal-type fuzzy set (usually we just use a pure trapezoidal fuzzy set as an output), and so on. We will see details in the following chapters.

Chapter 5

Fuzzy Knowledge Rediscovery in Fuzzy Rule Bases

In the two previous chapters, we have developed the theory of some novel techniques regarding effective fuzzy reasoning and genetic fuzzy neural networks. We will now examine and illustrate their applicability as well as their advantages. In the following Chapters, we will apply these techniques to various applications including function approximation, fuzzy control, compression and expansion of a fuzzy rule base, highly nonlinear system modeling and chaotic time series prediction.

In this chapter, the KRA given in Section 3.4 is used to rediscover fuzzy knowledge in fuzzy rule bases for two different applications which are

(1) an improvement of existing defuzzification techniques;

(2) an approximation of a nonlinear function.

5.1 Applicability of Various Defuzzification Techniques

The defuzzification technique plays an important role in fuzzy control and fuzzy decision-making since the performance of final crisp outputs directly affects the efficiency and performance of a fuzzy control-decision system. Based on output fuzzy sets generated by the Compositional Rule of Inference (CRI), various defuzzification techniques have been developed to try to find more reasonable crisp value to represent the output fuzzy sets [24,58,66,67,73,103,109,119]. If output fuzzy sets generated by the CRI are reasonable and meaningful, then related defuzzification techniques may be useful. So the CRI is the crucial basis on which the commonly used defuzzification techniques are built. In order to examine the correctness and efficiency of the CRI and the related defuzzification techniques, we use a simple fuzzy multiplication problem to analyze various defuzzification techniques such as (1) the Center of Gravity (COG) (also called Center of Area (COA)), (2) the Mean of Maximum (MOM), (3) the Center of Sums (COS)[24], (4) Center Average Defuzzifier (CAD) [103], (5) Modified Center Average Defuzzifier (MCAD) [103], (6) Yager's Level Set Method (Yager's LSM) [26], (7) Lin's Approximate Center of Area (Lin's ACOA) [72] and (8) our new NDM in Section 3.4. Actually, the COG and the MOM are two extreme cases of the BADD defuzzification method [109].

The two fuzzy rules for a given x, when $0.5 \leq x \leq 1.5$ according to [5] are given below,

Rule 1: IF x is $\tilde{0.5}$ THEN y is $\tilde{0.5}$,
Rule 2: IF x is $\tilde{1.0}$ THEN y is $\tilde{1.0}$,

where $\tilde{0.5}$ and $\tilde{1.0}$ are triangular fuzzy sets defined by below,

$$\tilde{0.5} = \begin{cases} 2x & \text{for } 0 \leq x \leq 0.5 \\ 2 - 2x & \text{for } 0.5 < x \leq 1.0, \end{cases} \tag{5.1}$$

$$\tilde{1.0} = \begin{cases} 2x - 1 & \text{for } 0.5 \leq x \leq 1.0 \\ 3 - 2x & \text{for } 1.0 < x \leq 1.5. \end{cases} \tag{5.2}$$

The question is "what is a reasonable value of y for $0 \leq x \leq 1.5$?". For this question, Bastian developed a technique to try to avoid the unnecessary nonlinearity generated by a conventional fuzzy reasoning technique [5]. However, Bastian's method cannot avoid the unnecessary nonlinearity completely. Intuitively, the given fuzzy rules imply a heuristic mapping function $y = x$. So we take $y = x$ as the ideal result based only on the two fuzzy rules.
(1) For $0 \leq x \leq 0.5$

In this case, only the first fuzzy rule is fired. Suppose we use the NFR with $\gamma^k = 0$ and Eqs. (3.50) and (3.52). The different results of fuzzy reasoning generated by the NFR and the sup-min CRI are shown in Figures 5.1(a) and 5.1(b), respectively.

By using the NDM, we have a normal fuzzy system such that

$$y = x \quad \text{for } 0 \leq x \leq 0.5. \tag{5.3}$$

By using the other above mentioned defuzzification methods, we have a fuzzy system such that

$$y = 0.5 \quad \text{for } 0 \leq x \leq 0.5. \tag{5.4}$$

Sec. 5.1. Applicability of Various Defuzzification Techniques

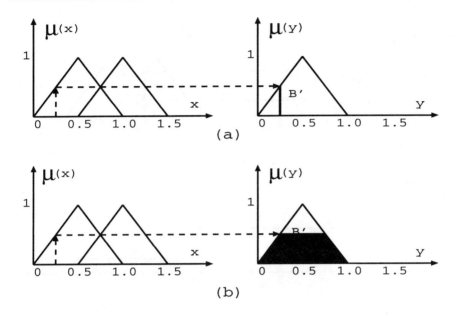

Figure 5.1. (a) The Fuzzy Reasoning Result According to The NFR. (b) The Fuzzy Reasoning Result According to The CRI.

Actually, any reasonable defuzzification method should get the conclusion $y = 0.5$ based on the output trapezoidal fuzzy set shown in Figure 5.1(b). Unfortunately, the output fuzzy set generated by the CRI is not reasonable, thus even very effective defuzzification methods cannot generate good crisp outputs.

(2) For $0.5 < x < 1.0$

In this case, the two fuzzy rules are fired. The different results generated by the NFR and the sup-min CRI are shown in Figures 5.2(a) and 5.2(b), respectively.

Based on the outputs in Figures 5.2(a) and 5.2(b), different defuzzification methods may generate different final crisp outputs as follows,
(2.1) the NDM

$$y = x \quad \text{for} \ \ 0.5 < x < 1.0, \tag{5.5}$$

(2.2) the COG

$$y = 0.75 + \frac{3 - 4x}{16x^2 - 24x + 4} \quad \text{for} \ \ 0.5 < x < 1.0, \tag{5.6}$$

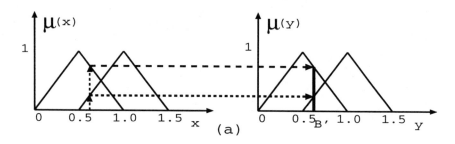

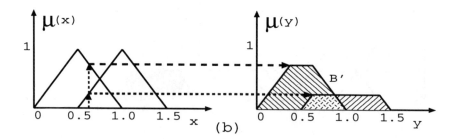

Figure 5.2. (a) The Fuzzy Reasoning Result According to The NFR; (b) The Fuzzy Reasoning Result According to The CRI.

(2.3) the COS

$$y = 0.75 + \frac{3 - 4x}{32x^2 - 48x + 12} \quad \text{for} \quad 0.5 < x < 1.0, \tag{5.7}$$

(2.4) the MOM

$$y = \begin{cases} 0.5 & \text{for } 0.5 \leq x < 0.75 \\ 0.75 & \text{for } x = 0.75 \\ 1.0 & \text{for } 0.75 < x \leq 1.0, \end{cases} \tag{5.8}$$

(2.5) the CAD

$$y = x \quad \text{for} \quad 0.5 < x < 1.0, \tag{5.9}$$

(2.6) the MCAD

$$y = x \quad \text{for} \quad 0.5 < x < 1.0, \tag{5.10}$$

(2.7) Yager's LSM

$$y = x \quad \text{for} \quad 0.5 < x < 1.0, \tag{5.11}$$

Sec. 5.1. Applicability of Various Defuzzification Techniques

(2.8) Lin's ACOA
$$y = x \quad \text{for} \quad 0.5 < x < 1.0. \tag{5.12}$$

(3) For $1.0 \leq x \leq 1.5$

Similarly, the crisp output y generated by the NDM is given by:
$$y = x \quad \text{for} \quad 1.0 \leq x \leq 1.5. \tag{5.13}$$

The crisp output y generated by the other defuzzification methods is given by
$$y = 1.0 \quad \text{for} \quad 1.0 \leq x \leq 1.5. \tag{5.14}$$

For clarity, all results are completely listed below and shown in Figure 5.3.

(a) the NDM
$$y = x \quad \text{for} \quad 0 \leq x \leq 1.5, \tag{5.15}$$

(b) the COG
$$y = \begin{cases} 0.5 & \text{for} \quad 0 \leq x \leq 0.5 \\ 0.75 + \frac{3-4x}{16x^2 - 24x + 4} & \text{for} \quad 0.5 < x < 1.0 \\ 1.0 & \text{for} \quad 1.0 \leq x \leq 1.5, \end{cases} \tag{5.16}$$

(c) the COS
$$y = \begin{cases} 0.5 & \text{for} \quad 0 \leq x \leq 0.5 \\ 0.75 + \frac{3-4x}{32x^2 - 48x + 12} & \text{for} \quad 0.5 < x < 1.0 \\ 1.0 & \text{for} \quad 1.0 \leq x \leq 1.5, \end{cases} \tag{5.17}$$

(d) the MOM
$$y = \begin{cases} 0.5 & \text{for} \quad 0 \leq x < 0.75 \\ 0.75 & \text{for} \quad x = 0.75 \\ 1.0 & \text{for} \quad 0.75 < x \leq 1.5, \end{cases} \tag{5.18}$$

(e) CAD, MCAD, Yager's LSM and Lin's ACOA
$$y = \begin{cases} 0.5 & \text{for} \quad 0 \leq x \leq 0.5 \\ x & \text{for} \quad 0.5 < x < 1.0 \\ 1.0 & \text{for} \quad 1.0 \leq x \leq 1.5. \end{cases} \tag{5.19}$$

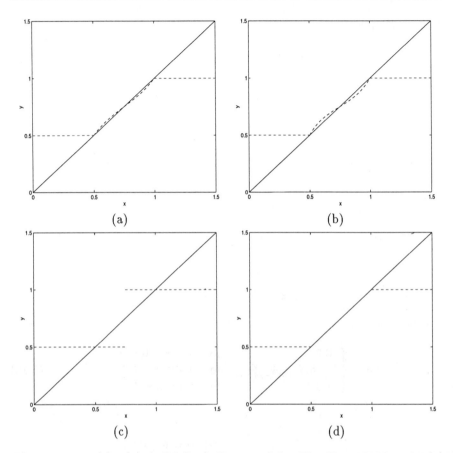

Figure 5.3. (a) $y(x)$ (solid line) Generated by The New NDM and $y(x)$ (dashed line) Generated by The COG. (b) $y(x)$ (solid line) Generated by The New NDM and $y(x)$ (dashed line) Generated by The COS. (c) $y(x)$ (solid line) Generated by The New NDM and $y(x)$ (dashed line) Generated by The MOM. (d) $y(x)$ (solid line) Generated by The New NDM and $y(x)$ (dashed line) Generated by The CAD, MCAD, Yager's LSM and Lin's ACOA.

Based on the simulations, we have found that the outputs of fuzzy reasoning and the final crisp outputs using the new normal reasoning methodology are more intuitive than those using the sup-min CRI since the given fuzzy rules imply the heuristic mapping function $y = x$ [5]. The reason for this is that the conventional CRI always takes non-primary fuzzy sets as fundamental elements

Sec. 5.2. Nonlinear Function Approximation

for fuzzy reasoning, it doesn't treat primary fuzzy subsets in a non-primary fuzzy set differently, thus many related defuzzification methods may generate the VICO problem such as $y = 0.5$ for $0 \leq x \leq 0.5$ and $y = 1$ for $1 \leq x \leq 1.5$. Actually, Bastian [5] only discussed the unnecessary nonlinearity problem, but didn't study the VICO problem. Importantly, our new NFR method has solved both the unnecessary nonlinearity problem and the VICO problem.

For completeness, a nonlinear fuzzy function approximation is discussed in the next section to compare the new normal reasoning methodology with the conventional CRI.

5.2 Nonlinear Function Approximation

A nonlinear function of x and y is defined as follows,

$$z = F(x, y) = (x - 2)^2 + (y - 3)^2. \tag{5.20}$$

A 5×5 fuzzy rule base is given in Table 5.1.

Table 5.1: 5×5 Fuzzy Rule Base for $F(x,y)$. (Row: x, Column: y).

	$\tilde{1}$	$\tilde{2}$	$\tilde{3}$	$\tilde{4}$	$\tilde{5}$
$\tilde{0}$	$\tilde{8}$	$\tilde{5}$	$\tilde{4}$	$\tilde{5}$	$\tilde{8}$
$\tilde{1}$	$\tilde{5}$	$\tilde{2}$	$\tilde{1}$	$\tilde{2}$	$\tilde{5}$
$\tilde{2}$	$\tilde{4}$	$\tilde{1}$	$\tilde{0}$	$\tilde{1}$	$\tilde{4}$
$\tilde{3}$	$\tilde{5}$	$\tilde{2}$	$\tilde{1}$	$\tilde{2}$	$\tilde{5}$
$\tilde{4}$	$\tilde{8}$	$\tilde{5}$	$\tilde{4}$	$\tilde{5}$	$\tilde{8}$

The 25 fuzzy rules in Table 5.1 are represented by

$$IF\ x\ is\ A_{ij}\ and\ y\ is\ B_{ij}\ THEN\ z\ is\ C_{ij}, \tag{5.21}$$

where fuzzy numbers $A_{ij} = (\widetilde{i-1})$ and $B_{ij} = \tilde{j}$ are defined by trapezoidal quaternions $(i-1, 0, i-1.75, i-0.25)$ and $(j, 0, j-0.75, j+0.75)$ for $i = 1, 2, ..., 5$ and $j = 1, 2, ..., 5$, respectively. Fuzzy numbers C_{ij} for $i = 1, 2, ..., 5$ and $j = 1, 2, ..., 5$ are defined by the trapezoidal quaternions shown in Table 5.2.

We can use the KRA to get heuristic parameters shown in Tables 5.3-5.6.

Table 5.2: Fuzzy Numbers C_{ij} for $i = 1, 2, ..., 5$ and $j = 1, 2, ..., 5$.

	$j=1$	$j=2$	$j=3$	$j=4$	$j=5$
$i=1$	$(8,0,7,9)$	$(5,0,4,6)$	$(4,0,3,5)$	$(5,0,4,6)$	$(8,0,7,9)$
$i=2$	$(5,0,4,6)$	$(2,0,1,3)$	$(1,0,0,2)$	$(2,0,1,3)$	$(5,0,4,6)$
$i=3$	$(4,0,3,5)$	$(1,0,0,2)$	$(0,0,-1,1)$	$(1,0,0,2)$	$(4,0,3,5)$
$i=4$	$(5,0,4,6)$	$(2,0,1,3)$	$(1,0,0,2)$	$(2,0,1,3)$	$(5,0,4,6)$
$i=5$	$(8,0,7,9)$	$(5,0,4,6)$	$(4,0,3,5)$	$(5,0,4,6)$	$(8,0,7,9)$

Table 5.3: Heuristic parameters w_{ij}^{aleft}.

-1	-1	-1	-1	-1
-1	-1	-1	-1	-1
-1	-1	-1	-1	-1
+1	+1	+1	+1	+1
+1	+1	+1	+1	+1

Table 5.4: Heuristic parameters w_{ij}^{aright}.

-1	-1	-1	-1	-1
-1	-1	-1	-1	-1
+1	+1	+1	+1	+1
+1	+1	+1	+1	+1
+1	+1	+1	+1	+1

Table 5.5: Heuristic parameters w_{ij}^{bleft}.

-1	-1	-1	+1	+1
-1	-1	-1	+1	+1
-1	-1	-1	+1	+1
-1	-1	-1	+1	+1
-1	-1	-1	+1	+1

Table 5.6: Heuristic parameters w_{ij}^{bright}.

-1	-1	+1	+1	+1
-1	-1	+1	+1	+1
-1	-1	+1	+1	+1
-1	-1	+1	+1	+1
-1	-1	+1	+1	+1

Sec. 5.2. Nonlinear Function Approximation

The normal fuzzy system $f(x,y)$ is given as follows,

$$f(x,y) = \frac{\sum_{i=1}^{5}\sum_{j=1}^{5}\phi_{ij}[\mu_{A_{ij}}(x)\mu_{B_{ij}}(y)]^{1-\frac{\gamma_{ij}}{2}}}{\sum_{i=1}^{5}\sum_{j=1}^{5}[\mu_{A_{ij}}(x)\mu_{B_{ij}}(y)]^{1-\frac{\gamma_{ij}}{2}}}, \quad (5.22)$$

where

$$\phi_{ij} = c_{ij} + \frac{\sigma_{ij}}{2}[\frac{w_{ij}^a(x-a_{ij})}{\alpha_{ij}} + \frac{w_{ij}^b(y-b_{ij})}{\beta_{ij}}], \quad (5.23)$$

where input centers $a_{ij} = i-1$ and $b_{ij} = j$, input widths $\alpha_{ij} = \beta_{ij} = 0.75$, c_{ij} are the centers of triangular membership functions of fuzzy sets defined in Table 5.2, output widths $\sigma_{ij} = 1$, and compensatory degrees $\gamma_{ij} = 0$.

Wang's fuzzy system [103] is described below,

$$g(x,y) = \frac{\sum_{i=1}^{5}\sum_{j=1}^{5} c_{ij}\mu_{A_{ij}}(x)\mu_{B_{ij}}(y)}{\sum_{i=1}^{5}\sum_{j=1}^{5} \mu_{A_{ij}}(x)\mu_{B_{ij}}(y)}. \quad (5.24)$$

Simulation results are given in Figures 5.4 and 5.5. Therefore, the simulations can support that the normal fuzzy system is more efficient than Wang's fuzzy system, and the novel fuzzy reasoning methodology can approach the nonlinear function (5.20) more accurately than the conventional one.

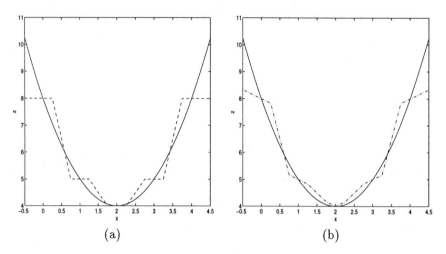

Figure 5.4. (a) $F(x,y)$ (solid line), $g(x,y)$ (dashed line) Generated by The Conventional Method When y=1. (b) $F(x,y)$ (solid line), $f(x,y)$ (dot-dashed line) Generated by The New Method When y=1.

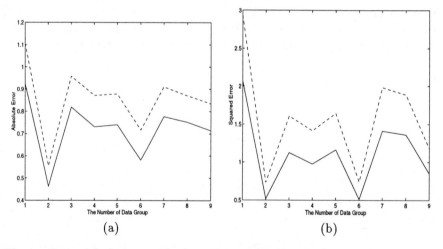

Figure 5.5. (a) Average Absolute Errors (dashed line) Generated by The Conventional Method and Average Absolute Errors (solid line) Generated by The New Method. (b) Average Squared Errors (dashed line) Generated by The Conventional Method and Average Squared Errors (solid line) Generated by The New Method.

Chapter 6

Fuzzy Cart-pole Balancing Control Systems

In this chapter, we continue to verify the efficiency of our normal fuzzy reasoning methodology and compensatory operations by doing simulations of a well known cart-pole balance system with crisp inputs and outputs. Furthermore, we verify the effectiveness of a CFNN which controls a complex cart-pole balance system with fuzzy inputs and outputs.

6.1 Cart-pole Balancing Fuzzy Control Systems

The dynamic equations of the cart-pole system shown in Figure 6.1 for an angle $\theta(t)$, an angular velocity $\dot{\theta}(t)$, a position $p(t)$ and a cart speed $\dot{p}(t)$ are given by

$$\theta(t+1) = \theta(t) + \Delta\dot{\theta}(t), \tag{6.1}$$

$$\dot{\theta}(t+1) = \dot{\theta}(t) + \Delta\frac{\phi(t) - cos\theta(t)[\psi(t) + m_p l(\dot{\theta}(t))^2 sin\theta(t)] - \frac{\mu_p m \dot{\theta}(t)}{m_p l}}{(4/3)ml - m_p l cos^2\theta(t)}, \tag{6.2}$$

$$p(t+1) = p(t) + \Delta\dot{p}(t), \tag{6.3}$$

$$\dot{p}(t+1) = \dot{p}(t) + \Delta\frac{\psi(t) + m_p l[(\dot{\theta}(t))^2 sin\theta(t) - \ddot{\theta}(t)cos\theta(t)]}{m}, \tag{6.4}$$

where $g = -9.8m/s^2$, $m = 1.1kg$, $m_p = 0.1kg$, $l = 0.5m$, $\mu_c = 0.0005$, $\mu_p = 0.000002$, $\Delta = 0.02$, $\phi(t) = mgsin\theta(t)$ and $\psi(t) = f(t) - \mu_c sgn(\dot{p}(t))$ [72].

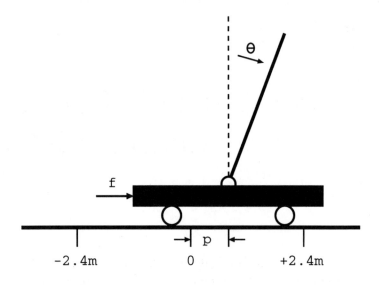

Figure 6.1. A cart-pole balancing system.

25 fuzzy rules are defined by

$$IF\ X_1\ is\ A_1^k\ and\ X_2\ is\ A_2^k \quad THEN\ Y\ is\ B^k, \qquad (6.5)$$

for $k = 0, 1, ..., 24$, where X_1 is a linguistic variable of the angle x_1 (i.e., $\theta(t)$), X_2 is a linguistic variable of the angular velocity x_2 (i.e., $\dot{\theta}(t)$), Y is a linguistic variable of the force y (i.e., f), relevant trapezoidal fuzzy sets A_1^k, A_2^k and B^k are defined by

$$\mu_{A_1^k}(x_1) = \begin{cases} \frac{x_1 - c_1^k}{a_1^k - b_1^k - c_1^k} & \text{for } c_1^k \leq x_1 < (a_1^k - b_1^k) \\ 1 & \text{for } (a_1^k - b_1^k) \leq x_1 \leq (a_1^k + b_1^k) \\ \frac{d_1^k - x_1}{d_1^k - a_1^k - b_1^k} & \text{for } (a_1^k + b_1^k) < x_1 \leq d_1^k \\ 0 & \text{otherwise,} \end{cases} \qquad (6.6)$$

$$\mu_{A_2^k}(x_2) = \begin{cases} \frac{x_2 - c_2^k}{a_2^k - b_2^k - c_2^k} & \text{for } c_2^k \leq x_2 < (a_2^k - b_2^k) \\ 1 & \text{for } (a_2^k - b_2^k) \leq x_2 \leq (a_2^k + b_2^k) \\ \frac{d_2^k - x_2}{d_2^k - a_2^k - b_2^k} & \text{for } (a_2^k + b_2^k) < x_2 \leq d_2^k \\ 0 & \text{otherwise,} \end{cases} \qquad (6.7)$$

Sec. 6.1. Cart-pole Balancing Fuzzy Control Systems

$$\mu_{B^k}(y) = \begin{cases} \frac{y-\eta^k}{\alpha^k-\beta^k-\eta^k} & \text{for } \eta^k \le y < (\alpha^k - \beta^k) \\ 1 & \text{for } (\alpha^k - \beta^k) \le y \le (\alpha^k + \beta^k) \\ \frac{\lambda^k-y}{\lambda^k-\alpha^k-\beta^k} & \text{for } (\alpha^k + \beta^k) < y \le \lambda^k \\ 0 & \text{otherwise,} \end{cases} \quad (6.8)$$

where $a_1^k = -\frac{\pi}{15} + \lfloor k/5 \rfloor \frac{\pi}{30}$, $b_1^k = \frac{\pi}{90}$, $c_1^k = a_1^k - \frac{\pi}{45}$, $d_1^k = a_1^k + \frac{\pi}{45}$, $a_2^k = -2 + k \bmod 5$, $b_2^k = 0.2$, $c_2^k = a_2^k - 0.75$, $d_2^k = a_2^k + 0.75$, $\alpha^k = -20$ for $k = 0, 1, 2, 5, 6$, $\alpha^k = -10$ for $k = 3, 7, 10, 11, 15$, $\alpha^k = 0$ for $k = 4, 8, 12, 16, 20$, $\alpha^k = 10$ for $k = 9, 13, 14, 17, 21$, $\alpha^k = 20$ $k = 18, 19, 22, 23, 24$, $\beta^k = 1.875$, $\eta^k = \alpha^k - 7.5$, $\lambda^k = \alpha^k + 7.5$.

For clarity, the above 5 × 5 fuzzy rule base is given in Table 6.1.

Table 6.1: A 5 × 5 fuzzy rule base.

	$\tilde{-2}$	$\tilde{-1}$	$\tilde{0}$	$\tilde{1}$	$\tilde{2}$
$-\frac{\pi}{15}$	$\tilde{-20}$	$\tilde{-20}$	$\tilde{-20}$	$\tilde{-10}$	$\tilde{0}$
$-\frac{\pi}{30}$	$\tilde{-20}$	$\tilde{-20}$	$\tilde{-10}$	$\tilde{0}$	$\tilde{10}$
0	$\tilde{-10}$	$\tilde{-10}$	$\tilde{0}$	$\tilde{10}$	$\tilde{10}$
$+\frac{\pi}{30}$	$\tilde{-10}$	$\tilde{0}$	$\tilde{10}$	$\tilde{20}$	$\tilde{20}$
$+\frac{\pi}{15}$	$\tilde{0}$	$\tilde{10}$	$\tilde{20}$	$\tilde{20}$	$\tilde{20}$

According to the normal fuzzy reasoning method, a normal fuzzy system $f(x_1, x_2)$ with compensatory degrees $\gamma^k = 0$ is given by

$$f(x_1, x_2) = \frac{\sum_{k=0}^{24} (\frac{f_1^k + f_2^k}{2}) \mu_{A_1^k}(x_1^k) \mu_{A_2^k}(x_2^k)}{\sum_{k=0}^{24} \mu_{A_1^k}(x_1^k) \mu_{A_2^k}(x_2^k)}, \quad (6.9)$$

where

$$f_1^k = \begin{cases} \frac{\alpha^k - \beta^k - \eta^k}{a_1^k - b_1^k - c_1^k}(x_1 - c_1^k) + \eta^k & \text{for } c_1^k \le x_1 < (a_1^k - b_1^k) \\ \frac{\beta^k}{b_1^k}(x_1 - a_1^k + b_1^k) + \alpha^k - \beta^k & \text{for } (a_1^k - b_1^k) \le x_1 \le (a_1^k + b_1^k) \\ \frac{\alpha^k + \beta^k - \lambda^k}{d_1^k - a_1^k - b_1^k}(d_1^k - x_1) + \lambda^k & \text{for } (a_1^k + b_1^k) < x_1 \le d_1^k \\ 0 & \text{otherwise,} \end{cases} \quad (6.10)$$

$$f_2^k = \begin{cases} \frac{\alpha^k - \beta^k - \eta^k}{a_2^k - b_2^k - c_2^k}(x_2 - c_2^k) + \eta^k & \text{for } c_2^k \le x_2 < (a_2^k - b_2^k) \\ \frac{\beta^k}{b_2^k}(x_2 - a_2^k + b_2^k) + \alpha^k - \beta^k & \text{for } (a_2^k - b_2^k) \le x_2 \le (a_2^k + b_2^k) \\ \frac{\alpha^k + \beta^k - \lambda^k}{d_2^k - a_2^k - b_2^k}(d_2^k - x_2) + \lambda^k & \text{for } (a_2^k + b_2^k) < x_2 \le d_2^k \\ 0 & \text{otherwise.} \end{cases} \quad (6.11)$$

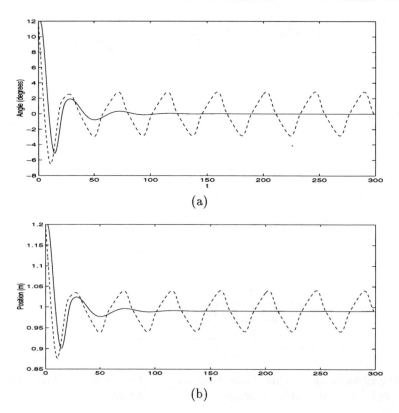

Figure 6.2. (a) Angles generated by our system (solid line) and by Wang's system (dashed line), (b) Positions generated by our system (solid line) and by Wang's system (dashed line) when (1) the initial angle is $12°$, (2) the initial angular velocity is $-2.0 radians/s$, (3) the initial cart position is $1.2m$, (4) the initial cart speed is $3m/s$.

For comparison, Wang's fuzzy system [103] is given by

$$g(x_1, x_2) = \frac{\sum_{k=0}^{24} \alpha^k \mu_{A_1^k}(x_1)\mu_{A_2^k}(x_2)}{\sum_{k=0}^{24} \mu_{A_1^k}(x_1)\mu_{A_2^k}(x_2)}. \quad (6.12)$$

Typical simulation results shown in Figure 6.2 can indicate Wang's fuzzy system is unstable and our fuzzy system is stable. Actually, the key reason is that our new fuzzy system doesn't have the VICO problem but Wang's fuzzy

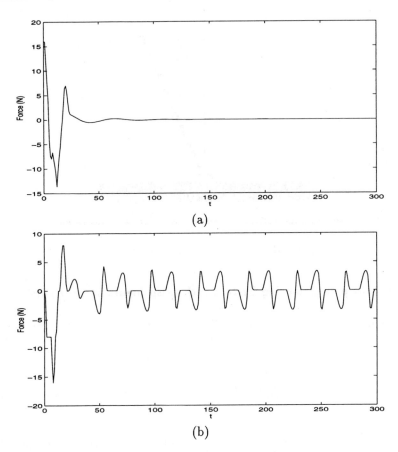

Figure 6.3. (a) Forces generated by our system, (b) Forces generated by Wang's system when (1) the initial angle is $12°$, (2) the initial angular velocity is $-2.0 radians/s$, (3) the initial cart position is $1.2m$, (4) the initial cart speed is $3m/s$.

system has it. In Figure 6.3(a), the force generated by our fuzzy system changes properly with the changing of angle and angular velocity. In Figure 6.3(b), a lot of constant forces (i.e., forces are 0N) generated by Wang's fuzzy system doesn't change reasonably with the changing of angle and angular velocity. Therefore, Wang's fuzzy system can not keep the cart-pole vertical. Finally, a lot of simulations under different initial states have indicated that our new fuzzy system is more robust than Wang's fuzzy system.

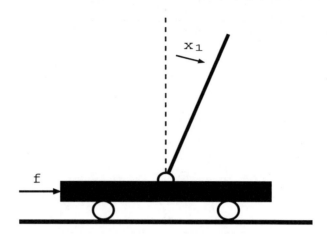

Figure 6.4. A cart-pole balancing system.

6.2 A Cart-pole Balancing System with Crisp Inputs and Outputs

The dynamic equations of the simplified cart-pole system shown in Figure 6.4 for the angle x_1 and the angular velocity $x_2 = \dot{x}_1$ are given by

$$\dot{x}_1 = x_2, \tag{6.13}$$

$$\dot{x}_2 = \frac{Mg\sin x_1 - \cos x_1[f - mlx_2^2 \sin x_1]}{l[\frac{4M}{3} - m\cos^2 x_1]}. \tag{6.14}$$

The above dynamic equations used by Wang [103] do not include equations for a cart's position. In order to compare our model with Wang's model under same conditions, we use the above simplified cart-pole dynamic system to verify our normal fuzzy system. We chose $g = 9.8 m/s^2$, $m = 0.1 kg$, $M = 1.1 kg$, and $l = 0.5m$. Since Wang used 25 fuzzy rules in the cart-pole system, we still use 25 fuzzy rules which are defined by

$$IF\ X_1\ is\ A_1^k\ and\ X_2\ is\ A_2^k \quad THEN\ F\ is\ B^k, \tag{6.15}$$

for $k = 0, 1, ..., 24$, where X_1 is a linguistic variable of the angle x_1, X_2 is a linguistic variable of the angular velocity x_2, F is a linguistic variable of the force f, relevant linguistic values A_1^k, A_2^k and B^k are defined by

$$A_1^k = \int_R \mu_{A_1^k}(x_1)/x_1 = \int_R exp[-\frac{(x_1 - \alpha(1 - \frac{1}{2}\lfloor \frac{k}{5} \rfloor))^2}{2(\beta_1 \alpha)^2}]/x_1, \tag{6.16}$$

Sec. 6.2. A Cart-pole Balancing System with Crisp Inputs and Outputs

$$A_2^k = \int_R \mu_{A_2^k}(x_2)/x_2 = \int_R exp[-\frac{(x_2 - \dot{\alpha}(1 - \frac{kmod5}{2}))^2}{2(\beta_2\dot{\alpha})^2}]/x_2, \quad (6.17)$$

$$B^k = \int_R \mu_{B^k}(f)/f = \int_R exp[-\frac{(f - \beta_3 b^k)^2}{2\sigma^2}]/f, \quad (6.18)$$

where α, $\dot{\alpha}$, σ, β_1, β_2 and β_3 are parameters to adjust the shapes of corresponding membership functions. $b^k = -8$ for $k = 0, 1, 2, 5, 6$; $b^k = -4$ for $k = 3, 7, 10, 11, 15$; $b^k = 0$ for $k = 4, 8, 12, 16, 20$; $b^k = 4$ for $k = 9, 13, 14, 17, 21$; $b^k = 8$ for $k = 18, 19, 22, 23, 24$.

A normal compensatory neuro-fuzzy system $f(x_1, x_2)$ directly based on given fuzzy rules,

$$f(x_1, x_2) = \frac{\sum_{k=0}^{24}(b^k + \frac{\eta^k}{2}\sum_{i=1}^{2}\frac{w_i^k(x_i - a_i^k)}{\sigma_i^k})[\prod_{i=1}^{2}\mu_{A_i^k}(x_i^k)]^{1-\frac{\gamma_k}{2}}}{\sum_{k=0}^{24}[\prod_{i=1}^{2}\mu_{A_i^k}(x_i^k)]^{1-\frac{\gamma_k}{2}}}, \quad (6.19)$$

where heuristic parameters w_i^k are defined below,

$$w_i^k = \begin{cases} wleft_i^k & \text{for } x_i \leq a_i^k \\ wright_i^k & \text{for } x_i > a_i^k. \end{cases} \quad (6.20)$$

For comparison, Wang's fuzzy system is given by

$$g(x_1, x_2) = \frac{\sum_{k=0}^{24} b^k \mu_{A_1^k}(x_1)\mu_{A_2^k}(x_2)}{\sum_{k=0}^{24} \mu_{A_1^k}(x_1)\mu_{A_2^k}(x_2)}. \quad (6.21)$$

Obviously, Wang's fuzzy system (6.21) is a simplified version of the normal fuzzy system (6.19) when $w_i^k = 0$ and $\gamma_k = 0$. . Actually, Wang's fuzzy system is the same as Lin's fuzzy system since the widths of all output membership functions are identical in this case. Additionally, the fuzzy system (6.21) is also the same as the fuzzy system using Yager's LSM [26] since all output membership functions are symmetric.

Now two typical cases are studied below.

CASE 1: $\gamma^k = 0$

We can easily get $w_i^k = 1$ by using the KRA, then make Takagi-Sugeno type fuzzy rules according to the high-level fuzzy rules given by (6.15), such that

$$IF\ X_1\ is\ A_1^k\ and\ X_2\ is\ A_2^k \quad THEN\ f^k = c_0^k + c_1^k x_1 + c_2^k x_2, \quad (6.22)$$

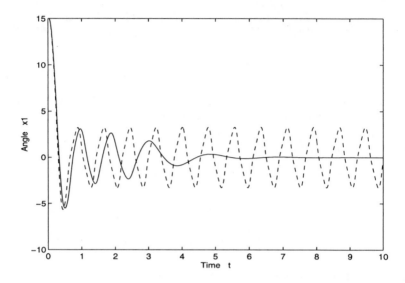

Figure 6.5. $x_1(t)$ (dashed line) generated by the conventional method and $x_1(t)$ (solid line) generated by the new method for $\beta_1 = \frac{1}{8}, \beta_2 = \frac{1}{8}, \beta_3 = 1, \sigma = \frac{3}{4}, \alpha = \frac{\pi}{18}, \dot{\alpha} = 2$. (Note: values of angles in this figure are in degrees)

where $c_0^k = b^k - \frac{\sigma}{2\beta_1}(1 - \frac{1}{2}\lfloor \frac{k}{5} \rfloor) - \frac{\sigma}{2\beta_2}(1 - \frac{k \bmod 5}{2})$, $c_1^k = \frac{\sigma}{2\beta_1 \alpha}$, $c_2^k = \frac{\sigma}{2\beta_2 \alpha}$. Since c_0^k, c_1^k and c_2^k are directly related to the parameters of membership functions (see Eqs. (6.16)-(6.18)), c_0^k, c_1^k and c_2^k have some corresponding physical meanings. Finally, the normal fuzzy cart-pole system is given by:

$$f_1(x_1, x_2) = \frac{\sum_{k=0}^{24}(c_0^k + c_1^k x_1 + c_2^k x_2) \mu_{A_1^k}(x_1) \mu_{A_2^k}(x_2)}{\sum_{k=0}^{24} \mu_{A_1^k}(x_1) \mu_{A_2^k}(x_2)}. \quad (6.23)$$

Importantly, this new method provides us with a simple way to construct the Takagi-Sugeno's fuzzy system, based directly on given fuzzy rules without using the conventional parameter estimation method based on sample data. Takagi-Sugeno's fuzzy system is only a special kind of normal fuzzy system with appropriate input and output membership functions. So the normal fuzzy reasoning methodology is a theoretical foundation of Takagi-Sugeno's fuzzy system.

By changing the initial angle and the parameters α, $\dot{\alpha}$, σ, β_1, β_2 and β_3, we can get simulation results, shown in Figures 6.5 and 6.6. For the small initial angle of 15°, the conventional fuzzy system cannot keep the cart-pole vertical

Sec. 6.2. A Cart-pole Balancing System with Crisp Inputs and Outputs

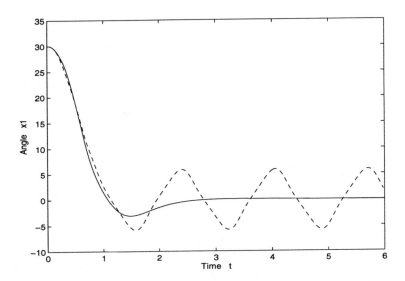

Figure 6.6. $x_1(t)$ (dashed line) generated by the conventional method and $x_1(t)$ (solid line) generated by the new method for $\beta_1 = \frac{1}{6}, \beta_2 = \frac{1}{8}, \beta_3 = 1, \sigma = 1, \alpha = \frac{\pi}{9}, \dot{\alpha} = 2$. (Note: values of angles in this figure are in degrees)

(see Figure 6.5). Intuitively, it is relatively easy for a fuzzy system to keep the cart-pole vertical for a small initial angle. So the conventional fuzzy system is not good enough. But the new normal fuzzy control system can effectively keep the cart-pole vertical for all given initial angles of 15° and 30° (see Figures 6.5 and 6.6). Therefore, the simulations can strongly support that the normal fuzzy cart-pole system is more efficient and more robust than the conventional fuzzy cart-pole system, and the novel fuzzy reasoning methodology is more reasonable and more powerful than the conventional one.

CASE 2: $\gamma = \gamma^k$ and $w_i^k = 0$

In this special case, we have a normal compensatory neuro-fuzzy system $f_2(x_1, x_2)$ directly based on given fuzzy rules as follows,

$$f_2(x_1, x_2) = \frac{\sum_{k=0}^{24} b^k [\prod_{i=1}^{2} \mu_{A_i^k}(x_i^k)]^{1-\frac{\gamma}{2}}}{\sum_{k=0}^{24} [\prod_{i=1}^{2} \mu_{A_i^k}(x_i^k)]^{1-\frac{\gamma}{2}}}, \qquad (6.24)$$

When $\gamma = 0$, the compensatory neuro-fuzzy system $f_2(x_1, x_2)$ becomes Wang's fuzzy system $g(x_1, x_2)$.

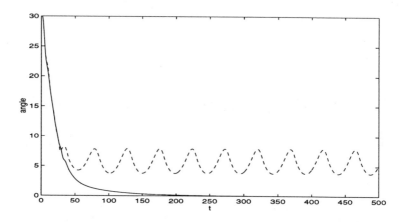

Figure 6.7. $x_1(t)$ (dashed line) generated by Wang's fuzzy system $g(x_1, x_2)$ and $x_1(t)$ (solid line) generated by the compensatory neuro-fuzzy system $f_2(x_1, x_2)$ for $\gamma = 0.9, \beta_1 = \frac{1}{8}, \beta_2 = \frac{1}{8}, \beta_3 = 6, \sigma = 4.5, \alpha = \frac{2\pi}{9}, \dot{\alpha} = 2$. (Note: values of angles in this figure are in degrees)

By changing the initial angles and γ, we can get typical simulation results, shown in Table 6.2 and Figures 6.7 and 6.8. The compensatory neuro-fuzzy system $f_2(x_1, x_2)$ can keep the cart-pole vertical for all given initial angles by choosing appropriate γ such as $\gamma = 0.9$ or $\gamma = 1$. But Wang's fuzzy system $g(x_1, x_2)$ cannot make the cart-pole system stable. Therefore, the simulations do indicate that a compensatory neuro-fuzzy system has a stronger ability to control a cart-pole balancing system than Wang's fuzzy system because a compensatory neuro-fuzzy system can adjust a compensatory degree. So the compensatory fuzzy reasoning makes a neuro-fuzzy system more adaptive.

Table 6.2: Stability of $f_2(x_1, x_2)$ under different γ. (S=Stable, US=UnStable.)

γ	0	0.1	0.2	0.3	0.4	0.5	0.6	0.7	0.8	0.9	1
$f_2(x_1, x_2)$	US	US	US	US	US	US	US	US	US	S	S

6.3 A Cart-pole Balancing System with Fuzzy Inputs and Outputs

The 25 fuzzy rules for a cart-pole balancing System are defined by

$$IF\ X1\ is\ A_1^k\ and\ X2\ is\ A_2^k \quad THEN\ F\ is\ B^k, \tag{6.25}$$

Sec. 6.3. A Cart-pole Balancing System with Fuzzy Inputs and Outputs

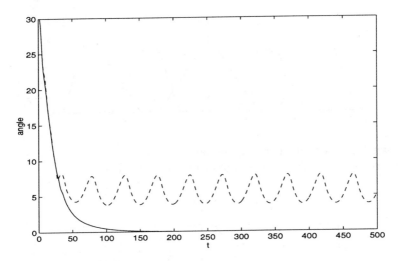

Figure 6.8. $x_1(t)$ (dashed line) generated by Wang's fuzzy system $g(x_1, x_2)$ and $x_1(t)$ (solid line) generated by the compensatory neuro-fuzzy system $f_2(x_1, x_2)$ for $\gamma = 1.0, \beta_1 = \frac{1}{8}, \beta_2 = \frac{1}{8}, \beta_3 = 6, \sigma = 4.5, \alpha = \frac{2\pi}{9}, \dot{\alpha} = 2$. (Note: values of angles in this figure are in degrees)

for $k = 0, 1, ..., 24$. Fuzzy linguistic values A_1^k, A_2^k and B^k are defined by Figures 6.9-6.11, respectively. A 5 × 5 fuzzy rule base for a cart-pole balancing system is given in Table 6.3.

Table 6.3: A 5 × 5 fuzzy rule base for a cart-pole balancing system.

	$\tilde{-2}$	$\tilde{-1}$	$\tilde{0}$	$\tilde{1}$	$\tilde{2}$
$-\frac{\pi}{3}$	$\tilde{-60}$	$\tilde{-60}$	$\tilde{-60}$	$\tilde{-24}$	$\tilde{0}$
$-\frac{\pi}{6}$	$\tilde{-60}$	$\tilde{-60}$	$\tilde{-24}$	$\tilde{0}$	$\tilde{24}$
0	$\tilde{-24}$	$\tilde{-24}$	$\tilde{0}$	$\tilde{24}$	$\tilde{24}$
$\frac{\pi}{6}$	$\tilde{-24}$	$\tilde{0}$	$\tilde{24}$	$\tilde{60}$	$\tilde{60}$
$\frac{\pi}{3}$	$\tilde{0}$	$\tilde{24}$	$\tilde{60}$	$\tilde{60}$	$\tilde{60}$

For comparison, Wang's fuzzy system is given by

$$f = \frac{\sum_{k=0}^{24} b^k \mu_{A_1^k}(x_1) \mu_{A_2^k}(x_2)}{\sum_{k=0}^{24} \mu_{A_1^k}(x_1) \mu_{A_2^k}(x_2)}. \qquad (6.26)$$

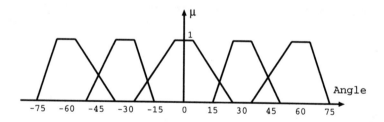

Figure 6.9. Asymmetric trapezoidal fuzzy sets of angles for a cart-pole balancing system. (Note: values of angle are in degrees)

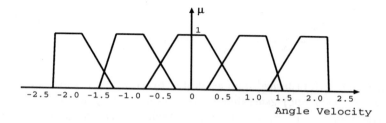

Figure 6.10. Asymmetric trapezoidal fuzzy sets of angle velocities for a cart-pole balancing system. (Note: values of angular velocity are in radians/second)

Conventional neural networks and fuzzy systems control a cart-pole system based on crisp inputs and outputs such as "if *angle* is $\frac{\pi}{4}$ and *angle velocity* is -1.5 then *force* will be 7.85 N". Obviously, these conventional techniques use a *singleton fuzzifier* to fuzzify a crisp value. However, some noise usually

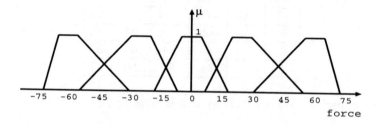

Figure 6.11. Asymmetric trapezoidal fuzzy sets of forces for a cart-pole balancing system. (Note: values of force are in Ns)

Sec. 6.3. A Cart-pole Balancing System with Fuzzy Inputs and Outputs

exists around a crisp value in a real application, so we have to consider how to make a robust neuro-fuzzy system to deal with noise around a crisp point. In other words, we have to use a *non-singleton fuzzifier* such as a trapezoidal fuzzy set and a Gaussian fuzzy set to deal with noise. A much more complex problem is how to control a cart-pole if both inputs and outputs are fuzzy. For example, "if *angle* is *around* $\frac{\pi}{4}$ and *angle velocity* is *about* -1.5 then *force* will be *around* 7.85 N", and finally the fuzzy concept *around* 7.85 N is defuzzified to a crisp value to control the cart-pole system. In order to simulate noise around a crisp value, commonly used crisp *angle*, *angle velocity* and *force* are randomly fuzzified to trapezoidal fuzzy sets because a device cannot absolutely measure not only input values but also output values. Therefore, we add noise to inputs and an output to verify the robustness and stability of fuzzy systems. For completeness, we use both a *singleton fuzzifier* and a *non-singleton fuzzifier* to do simulations. By changing the initial fuzzy angle and fuzzy angle velocity and adding random fuzzy noise to two inputs (i.e., *angle* and *angle velocity*) and one output (i.e., *force*), we can get typical simulation results, shown in Figure 6.12 (without fuzzy noise) and Figure 6.13 (with fuzzy noise). The new CFNN can effectively keep the cart-pole vertical for all given initial angles, but Wang's fuzzy system fails to do so. Therefore, the CFNN is more robust than Wang's fuzzy system.

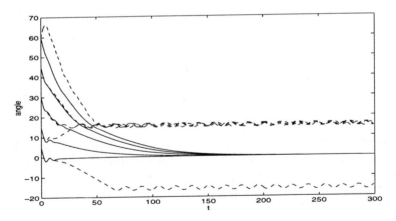

Figure 6.12. Angles (solid lines) generated by our model and angles (dashed lines) generated by Wang's model under different initial angles when no fuzzy noise inputs and outputs are applied to a cart-pole balancing system. (Note: values of angles in this figure are in degrees, and initial angles are 5°, 15°, 30°, 45°, 60°)

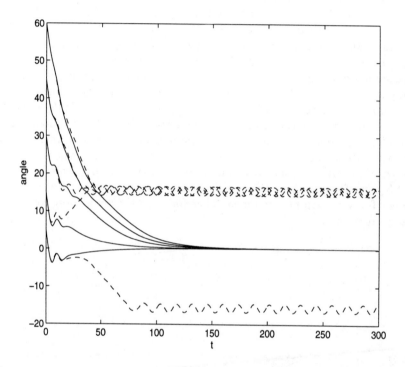

Figure 6.13. Angles (solid lines) generated by our model and angles (dashed lines) generated by Wang's model under different initial angles when fuzzy noise inputs and outputs are applied to a cart-pole balancing system. (Note: values of angles in this figure are in degrees, and initial angles are 5°, 15°, 30°, 45°, 60°)

Chapter 7

Fuzzy Knowledge Compression and Expansion

In the previous chapter, the simulations have indicated that the new normal fuzzy system can control a complex cart-pole balance system. Now we apply a CFNN to (1) compression of a fuzzy rule base and (2) expansion of a fuzzy rule base.

7.1 Compression of Fuzzy Rule Bases

In many applications, the complexity of fuzzy rule bases is a central problem when dealing with a large number of linguistic variables and linguistic values. How to effectively solve the problem of high dimensionality of fuzzy rule bases is very important in real applications of fuzzy logic. Babuska et al. [4] developed two related yet distinct techniques for simplification and reduction of fuzzy models: rule base simplification and linguistic approximation. Both models use a similarity measure to assess the similarity of the fuzzy sets in the rule base. In addition, Krone [63] investigated rule reduction methods which allow the reduction of the number of rules and the number of conflicts in the rule base. However, an important problem which is "how to compress N fuzzy rules to arbitrary M ($1 \leq M < N$) fuzzy rules?" is still unsolved. Here, a CFNN is used to directly compress a big fuzzy rule base with N fuzzy rules to a small one with M ($1 \leq M < N$) fuzzy rules.

The compression method using a CFNN is given below:

Suppose: an original fuzzy rule base has N fuzzy rules, we need to compress it to a compressed fuzzy rule base with M ($1 \leq M < N$) fuzzy rules.

Begin

Step 1: Transform trapezoidal-type fuzzy sets in the original fuzzy rule base to corresponding fuzzy feature quaternions such as fuzzy information capacity quaternions, trapezoidal quaternions for trapezoidal fuzzy sets and Gaussian quaternions for trapezoidal-Gaussian fuzzy sets. In other words, a user has to design mapping functions for the first layer of a CFNN. Then use the fuzzy feature quaternions of fuzzy sets in IF-part and those in THEN-part of fuzzy rules in the fuzzy rule base as inputs and outputs of a CFNN, respectively.

Step 2: Train the CFNN based on the generated fuzzy feature inputs and outputs to meet required errors.

Step 3: Input fuzzy feature quaternions of fuzzy sets of IF part of M compressed fuzzy rules to the trained CFNN, then the CFNN can generate corresponding fuzzy feature quaternions of fuzzy sets of THEN part of M compressed fuzzy rules. Finally, M compressed fuzzy rules have been constructed.

Step 4: Test the compressed fuzzy rule base:

IF its performance is acceptable,

THEN stop.

ELSE use the trained CFNN to generate sufficient K $(K > N)$ fuzzy outputs for fuzzy inputs, then transform input fuzzy sets and output fuzzy sets to input fuzzy feature quaternions and output ones, respectively. Goto *Step 2*.
End.

For example, we already have the 5×5 fuzzy rule base for a cart-pole balancing system (see Table 6.1). Now we need to compress it to a 3×3 compressed fuzzy rule base. For clarity, we show how to use the compression method to compress the 5×5 fuzzy rule base to a 3×3 compressed one step by step.

Step 1: Transform all trapezoidal fuzzy sets in Table 6.1 to corresponding trapezoidal quaternions according to Definition 4.1 in Section 4.4. For instance, a trapezoidal fuzzy set $\frac{\tilde{\pi}}{15}$ is transformed to a trapezoidal quaternion $(\frac{\tilde{\pi}}{15}, \frac{\tilde{\pi}}{90}, \frac{2\pi}{45}, \frac{4\pi}{45})$. Finally, all fuzzy sets in Table 7.1 can be represented by corresponding trapezoidal quaternions.

Step 2: Use generated trapezoidal quaternions of fuzzy sets in IF-part and those in THEN-part of 25 fuzzy rules in Table 6.1 as inputs and outputs of a CFNN, respectively. Then train FNNKD-1, FNNKD-2, FNNKD-3 and FNNKD-4 separately.

Sec. 7.1. Compression of Fuzzy Rule Bases

Step 3: In Table 7.1, $-\tilde{\frac{\pi}{20}}$, $\tilde{0}$ and $\tilde{\frac{\pi}{20}}$ are defined by $(-\frac{\pi}{20}, \frac{\pi}{90}, -\frac{4\pi}{45}, -\frac{\pi}{90})$, $(0, \frac{\pi}{90}, -\frac{7\pi}{180}, \frac{7\pi}{180})$ and $(\frac{\pi}{20}, \frac{\pi}{90}, \frac{\pi}{90}, \frac{4\pi}{45})$, respectively. $-\tilde{1}.5$, $\tilde{0}$ and $\tilde{1}.5$ are defined by trapezoidal quaternions $(-1.5, 0.2, -2.75, -0.25)$, $(0, 0.2, -1.25, 1.25)$ and $(1.5, 0.2, 0.25, 2.75)$, respectively.

Table 7.1: An Incomplete 3 × 3 Compressed Fuzzy Rule Base.

	$-\tilde{1}.5$	$\tilde{0}$	$\tilde{1}.5$
$-\frac{\pi}{20}$			
0			
$\frac{\pi}{20}$			

Use the trapezoidal quaternions of input fuzzy sets of the 9 compressed fuzzy rules in Table 7.1 to train CFNN, then the trained CFNN can generate corresponding trapezoidal quaternions of output fuzzy sets of the 9 compressed fuzzy rules. Finally, a complete 3 × 3 compressed fuzzy rule base is constructed in Table 7.2.

The fuzzy sets of forces in Table 7.2 are defined in Table 7.3.

Table 7.2: A Complete 3 × 3 Compressed Fuzzy Rule Base.

	$-\tilde{1}.5$	$\tilde{0}$	$\tilde{1}.5$
$-\frac{\pi}{20}$	$-\tilde{20}$	$-\tilde{15}$	$\tilde{0}$
0	$-\tilde{10}$	$\tilde{0}$	$\tilde{10}$
$\frac{\pi}{20}$	$\tilde{0}$	$\tilde{15}$	$\tilde{20}$

Table 7.3: The Fuzzy Sets of Forces in Table 7.2.

	$-\tilde{1}.5$	$\tilde{0}$	$\tilde{1}.5$
$-\frac{\pi}{20}$	$(-20, 1.87, -27.5, -12.5)$	$(-15, 1.87, -22.5, -7.5)$	$(0, 1.87, -7.49, 7.49)$
0	$(-10, 1.88, -17.5, -2.5)$	$(0, 1.87, -7.5, 7.5)$	$(10, 1.87, 2.51, 17.5)$
$\frac{\pi}{20}$	$(0, 1.88, -7.5, 7.5)$	$(15, 1.88, 7.49, 22.5)$	$(20, 1.88, 12.5, 27.5)$

Step 4: A lot of simulations to test the compressed fuzzy rule base can indicate that Wang's system is unstable and our system is stable. Typical simulation results are given in Figure 7.1. Finally, the performance of our system based on the compressed fuzzy rule base is acceptable.

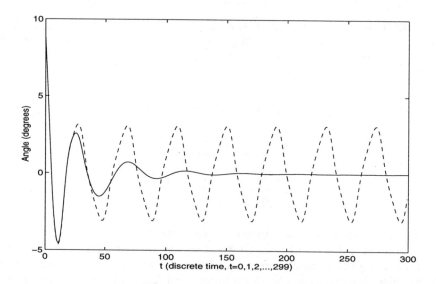

Figure 7.1. Angle (dashed line) Generated by Wang's Method and Angle (solid line) Generated by Our New Method Based on The 3 × 3 Compressed Fuzzy Rule Base When (1) The Initial Angle is 9°, (2) The Initial Angular Velocity is $-1.5 radians/s$, (3) The Initial Cart Position is $1m$, (4) The Initial Cart Speed is $3m/s$.

7.2 Expansion of Fuzzy Rule Bases

A sparse fuzzy rule base is a fuzzy rule base in which no rule will be fired for legitimate values. For example, a 3×3 sparse fuzzy rule base for a cart-pole balancing system is given in Table 7.4. In Table 7.4, the fuzzy sets $-\frac{\tilde{\pi}}{20}$, $\tilde{0}$ and $\frac{\tilde{\pi}}{20}$ are defined by trapezoidal quaternions $(-\frac{\pi}{20}, \frac{\pi}{90}, -\frac{\pi}{15}, -\frac{\pi}{30})$, $(0, \frac{\pi}{90}, -\frac{\pi}{60}, -\frac{\pi}{60})$ and $(\frac{\pi}{20}, \frac{\pi}{90}, \frac{\pi}{30}, \frac{\pi}{15})$, respectively. The fuzzy sets $-\tilde{1.5}$, $\tilde{0}$ and $\tilde{1.5}$ are defined by $(-1.5, 0.2, -2, -1)$, $(0, 0.2, -0.5, 0.5)$ and $(1.5, 0.2, 1, 2)$, respectively. The fuzzy sets of forces are defined by the trapezoidal quaternions shown in Table 7.5. Since the sparse 3×3 fuzzy rule base has special regions such as $(-6°, -3°)$ and $(3°, 6°)$ in Figure 7.2(a) and $(-1, -0.5)$ and $(0.5, 1)$ in Figure 7.2(b) where no rule will be fired, we have to develop a method to solve the sparse problem.

To expand a sparse fuzzy rule base to a non-sparse one, we have to define new input fuzzy sets to cover invalid regions in the sparse fuzzy rule base, and then get new corresponding output fuzzy sets in the non-sparse fuzzy rule base.

Sec. 7.2. Expansion of Fuzzy Rule Bases

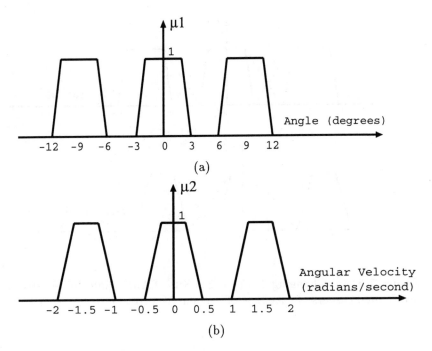

Figure 7.2. (a) Sparse Fuzzy Sets of Angles. (b) Sparse Fuzzy Sets of Angular Velocities.

Table 7.4: A Sparse 3 × 3 Fuzzy Rule Base.

	$-\tilde{1.5}$	$\tilde{0}$	$\tilde{1.5}$
$-\frac{\pi}{20}$	$-\tilde{20}$	$-\tilde{15}$	$\tilde{0}$
$\tilde{0}$	$-\tilde{10}$	$\tilde{0}$	$\tilde{10}$
$\frac{\pi}{20}$	$\tilde{0}$	$\tilde{15}$	$\tilde{20}$

Table 7.5: The Fuzzy Sets of Forces in Table 7.4.

	$-\tilde{1.5}$	$\tilde{0}$	$\tilde{1.5}$
$-\frac{\pi}{20}$	$(-20, 1.875, -27.5, -12.5)$	$(-15, 1.875, -22.5, -7.5)$	$(0, 1.875, -7.5, 7.5)$
$\tilde{0}$	$(-10, 1.875, -17.5, 2.5)$	$(0, 1.875, -7.5, 7.5)$	$(10, 1.875, 2.5, 17.5)$
$\frac{\pi}{20}$	$(0, 1.875, -7.5, 7.5)$	$(15, 1.875, 7.5, 22.5)$	$(20, 1.875, 12.5, 27.5)$

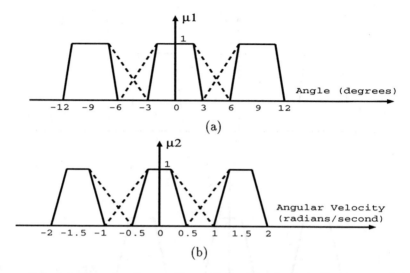

Figure 7.3. (a) Changing Sparse Fuzzy Sets of Angles into Non-sparse Ones. (b) Changing Sparse Fuzzy Sets of Angular Velocities into Non-sparse Ones.

For clarity, the expansion method is described below,

Step 1: Transform all trapezoidal fuzzy sets in Table 7.4 to corresponding trapezoidal quaternions according to Definition 4.1 in Section 4.4. Then use generated trapezoidal quaternions of fuzzy sets in IF-part and those in THEN-part of fuzzy rules in the original fuzzy rule base as inputs and outputs of a CFNN, respectively. Finally train FNNKD-1, FNNKD-2, FNNKD-3 and FNNKD-4 separately to meet required errors. Actually, a simple method to train FNNKD-1 is (1) redefine fuzzy sets near those invalid regions in an original sparse fuzzy rule base (e.g., use the dashed lines to redefine fuzzy sets to cover invalid regions in Figures 7.3(a) and 7.3(b)), (2) use new fuzzy sets to train FNNKD-1. (Note: we use the simple method in simulations.)

Step 2: New $-\frac{\tilde{\pi}}{40}$, $\frac{\tilde{\pi}}{40}$, $-\tilde{0.75}$ and $\tilde{0.75}$ are defined by trapezoidal quaternions $(-\frac{\tilde{\pi}}{40}, \frac{\tilde{\pi}}{90}, -\frac{\tilde{\pi}}{24}, -\frac{\tilde{\pi}}{120})$, $(\frac{\tilde{\pi}}{40}, \frac{\tilde{\pi}}{90}, -\frac{\tilde{\pi}}{120}, -\frac{\tilde{\pi}}{24})$, $(-0.75, 0.2, -1.25, -0.25)$ and $(0.75, 0.2, 0.25, 1.25)$, respectively, to cover the invalid regions shown in Figures 7.3(a) and 7.3(b). At this step, we have expanded the 3 × 3 sparse fuzzy rule base in Table 7.4 to an incomplete 5 × 5 expanded fuzzy rule base shown in Table 7.6.

Sec. 7.2. Expansion of Fuzzy Rule Bases

Table 7.6: An Incomplete 5 × 5 Expanded Fuzzy Rule Base.

	$-\dot{1}.5$	$-\dot{0}.75$	$\dot{0}$	$\dot{0}.75$	$\dot{1}.5$
$-\frac{\pi}{20}$	$-\dot{2}0$		$-\dot{1}5$		$\dot{0}$
$-\frac{\pi}{40}$					
$\dot{0}$	$-\dot{1}0$		$\dot{0}$		$\dot{1}0$
$\frac{\pi}{40}$					
$\frac{\pi}{20}$	$\dot{0}$		$\dot{1}5$		$\dot{2}0$

Step 3: Use the trained CFNN to generate 16 unknown output fuzzy sets in Table 7.6. Finally, a complete 5 × 5 expanded fuzzy rule base is given in Table 7.7, and fuzzy sets of forces in Table 7.7 are given in Table 7.8.

Table 7.7: A Complete 5 × 5 Expanded Fuzzy Rule Base.

	$-\dot{1}.5$	$-\dot{0}.75$	$\dot{0}$	$\dot{0}.75$	$\dot{1}.5$
$-\frac{\pi}{20}$	$-\dot{2}0$	$-\dot{1}7.5$	$-\dot{1}5$	$-\dot{7}.5$	$\dot{0}$
$-\frac{\pi}{40}$	$-\dot{1}5$	$-\dot{1}1.25$	$-\dot{7}.5$	$-\dot{1}.25$	$\dot{5}$
$\dot{0}$	$-\dot{1}0$	$-\dot{5}$	$\dot{0}$	$\dot{5}$	$\dot{1}0$
$\frac{\pi}{40}$	$-\dot{5}$	$\dot{1}.25$	$\dot{7}.5$	$\dot{1}1.25$	$\dot{1}5$
$\frac{\pi}{20}$	$\dot{0}$	$\dot{7}.5$	$\dot{1}5$	$\dot{1}7.5$	$\dot{2}0$

Table 7.8: The Fuzzy Sets of Forces in Table 7.7.

Angle	Angular Velocity	Force
$-\frac{\pi}{20}$	$-\dot{1}.5$	$(-20, 1.875, -27.5, -12.5)$
$-\frac{\pi}{20}$	$-\dot{0}.75$	$(-17.5, 1.876, -25.0, -9.998)$
$-\frac{\pi}{20}$	$\dot{0}$	$(-15, 1.875, -22.5, -7.5)$
$-\frac{\pi}{20}$	$\dot{0}.75$	$(-7.5, 1.877, -15.0, 0.004)$
$-\frac{\pi}{20}$	$\dot{1}.5$	$(0, 1.875, -7.5, 7.5)$
$-\frac{\pi}{40}$	$-\dot{1}.5$	$(-15, 1.871, -22.50, -7.502)$
$-\frac{\pi}{40}$	$-\dot{0}.75$	$(-11.25, 1.877, -18.75, -3.748)$
$-\frac{\pi}{40}$	$\dot{0}$	$(-7.5, 1.875, -15.0, 0.0)$
$-\frac{\pi}{40}$	$\dot{0}.75$	$(-1.25, 1.878, -8.756, 6.256)$
$-\frac{\pi}{40}$	$\dot{1}.5$	$(5, 1.875, -2.5, 12.5)$
$\dot{0}$	$-\dot{1}.5$	$(-10, 1.875, -17.5, 2.5)$
$\dot{0}$	$-\dot{0}.75$	$(-5, 1.877, -12.5, 2.504)$
$\dot{0}$	$\dot{0}$	$(0, 1.875, -7.5, 7.5)$
$\dot{0}$	$\dot{0}.75$	$(5, 1.878, -2.506, 12.51)$
$\dot{0}$	$\dot{1}.5$	$(10, 1.875, 2.5, 17.5)$
$\frac{\pi}{40}$	$-\dot{1}.5$	$(-5, 1.877, -12.5, 2.504)$
$\frac{\pi}{40}$	$-\dot{0}.75$	$(1.25, 1.877, -6.254, 8.754)$
$\frac{\pi}{40}$	$\dot{0}$	$(7.5, 1.875, 0.0, 15.0)$
$\frac{\pi}{40}$	$\dot{0}.75$	$(11.25, 1.879, 3.7420, 18.76)$
$\frac{\pi}{40}$	$\dot{1}.5$	$(15, 1.875, 7.5, 22.5)$
$\frac{\pi}{20}$	$-\dot{1}.5$	$(0, 1.875, -7.5, 7.5)$
$\frac{\pi}{20}$	$-\dot{0}.75$	$(7.5, 1.877, -0.004, 15.0)$
$\frac{\pi}{20}$	$\dot{0}$	$(15, 1.875, 7.5, 22.5)$
$\frac{\pi}{20}$	$\dot{0}.75$	$(17.5, 1.879, 9.992, 25.01)$
$\frac{\pi}{20}$	$\dot{1}.5$	$(20, 1.875, 12.5, 27.5)$

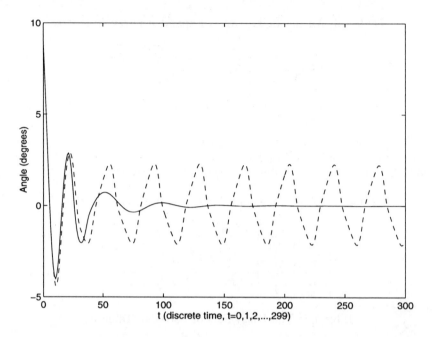

Figure 7.4. Angle (dashed line) Generated by Wang's Method and Angle (solid line) Generated by Our New Method Based on The 5 × 5 Expanded Fuzzy Rule Base When (1) The Initial Angle is 9°, (2) The Initial Angular Velocity is $-1.5 radians/s$, (3) The Initial Cart Position is $1m$, (4) The Initial Cart Speed is $3m/s$.

At this step, the original 3 × 3 sparse fuzzy rule base has been expanded to a 5 × 5 non-sparse fuzzy rule base. The typical simulations testing the 5 × 5 expanded fuzzy rule base are shown in Figure 7.4. Simulations have indicated that our model is more stable than Wang's model.

In summary, according to the new normal fuzzy reasoning methodology based on heuristic primary fuzzy sets, a Fuzzy Neural Network with Knowledge Discovery (FNNKD) is designed to perform an adaptive compensatory fuzzy reasoning. Since weights in a conventional crisp neural network and a fuzzy-operation-oriented neural network have no explicit physical meaning, they are not convenient to extract fuzzy rules from data. In order to overcome the weaknesses of conventional crisp neural networks and fuzzy-operation-oriented neural networks, we have developed a general fuzzy-reasoning-oriented fuzzy neural network called a Crisp-Fuzzy Neural Network (CFNN) which is able to extract high-level knowledge such as fuzzy IF-THEN rules from either crisp data or fuzzy data. A CFNN can effectively compress a 5 × 5 fuzzy IF-THEN rule base of a cart-pole balancing system to a 3 × 3 fuzzy rule base, and can expand an invalid sparse 3 × 3 fuzzy IF-THEN rule base to a valid 5 × 5 fuzzy rule base. Therefore, a CFNN is a useful soft computing system with both *linguistic-words*-level fuzzy reasoning and *numerical-data*-level information processing.

Chapter 8

Highly Nonlinear System Modeling and Prediction

In this chapter, we use three frequently used benchmarks to compare our new fuzzy neural networks with other well known systems in order to verify our new models objectively under various circumstances.

8.1 Nonlinear Function Prediction

A nonlinear function used by Takagi, et al. [94], Sugeno, et al. [92], Kondo [59] and Jang [43,44] is given by

$$f(x, y, z) = (1 + x^{0.5} + y^{-1} + z^{-1.5})^2, \qquad (8.1)$$

with the average percentage relative error defined as follows:

$$APE = \frac{1}{P} \sum_{i=1}^{P} \frac{|T(i) - O(i)|}{|T(i)|} * 100\%. \qquad (8.2)$$

Jang used 216 training data and 125 test data to verify the ANFIS rules (which is a special function in the Fuzzy Logic Toolbox in MATLAB) [43,44]. Unfortunately, the other approaches [59,92,94] used the 20 training data and 20 checking data to examine the performance, therefore no conclusive comments were made in [43] since different data were used. In order to completely compare our approach with the other ones, we will use not only both the above groups of data, but also three other groups of data to effectively analyze actual performance of the different approaches.

Sec. 8.1. Nonlinear Function Prediction

Case 1: 20 training data and 20 checking data

At first, we use our FNNKD and Jang's ANFIS provided in the Fuzzy Toolbox of MATLAB [44] to train 8 fuzzy rules based on the 20 training data used in [59,92,94], and then predict the values for the other 20 checking data.

We trained a 3-input-1-output FNNKD based on the 20 training data. After training the FNNKD (Note: here we don't train γ^k and set $\gamma^k = 0$), the FNNKD's training APE and checking APE are 1.83% and 2.28%, respectively.

It took only 2 epochs to train the ANFIS because after 2 epochs the RMSE (Root Mean Squared Error) was 0.00001 and the APE was 0.0001%. Interestingly, although the trained ANFIS is near perfect for the 20 training data, the RMSE and the APE of predicted data generated by it are 2.816 and 13.7%, respectively. The results are listed in Table 8.1 (Note: the other data come from [43]). In this case, the FNNKD is better than the ANFIS (Note: the number of parameters of the FNNKD is larger than that of the ANFIS).

Table 8.1: Comparisons with Earlier Works for Cases 1 and 2.

Model	$APE_{trn}(\%)$	$APE_{chk}(\%)$	Parameter No.	N_{trn}	N_{chk}
FNNKD	2.47	3.33	112	216	125
ANFIS	0.043	1.066	50	216	125
FNNKD	1.83	2.28	112	20	20
ANFIS	0.0001	13.7	50	20	20
GMDH model	4.7	5.7	–	20	20
Fuzzy model 1	1.5	2.1	22	20	20
Fuzzy model 2	0.59	3.4	32	20	20

Case 2: 216 training data and 125 checking data

The 216 training data and 125 checking data were sampled uniformly from the input ranges $[1,6] \times [1,6] \times [1,6]$ and $[1.5, 5.5] \times [1.5, 5.5] \times [1.5, 5.5]$, respectively. The simulation results generated by the ANFIS and the FNNKD are also listed in Table 8.1. In this case, the ANFIS's APE is 1.066%, and the FNNKD's APE is 3.33%. Since different training data and checking data may result in different simulation performance of the methods, we have to use different training data and checking data to examine the performance of the ANFIS and the FNNKD. Because both the ANFIS and the FNNKD try to minimize the RMSE of training data and then predict checking data, we use RMSE of predicted data to compare the ANFIS with the FNNKD in the following cases.

Case 3: 125 training data and 64 checking data

The 125 training data and 64 checking data were sampled uniformly from the input ranges $[1.5, 5.5] \times [1.5, 5.5] \times [1.5, 5.5]$ and $[1, 4] \times [1, 4] \times [1, 4]$, respectively.

Case 4: 64 training data and 27 checking data

The 64 training data and 27 checking data were sampled uniformly from the input ranges $[1, 4] \times [1, 4] \times [1, 4]$ and $[1.5, 3.5] \times [1.5, 3.5] \times [1.5, 3.5]$, respectively.

Case 5: 27 training data and 8 checking data

The 27 training data and 8 checking data were sampled uniformly from the input range $[1.5, 3.5] \times [1.5, 3.5] \times [1.5, 3.5]$ and $[1, 2] \times [1, 2] \times [1, 2]$, respectively.

The training speeds of the ANFIS and the FNNKD are listed in Table 8.2. Table 8.2 can indicate the training speed of the FNNKD is about 10 times faster than that of the ANFIS. The RMSEs of predicted data generated by the ANFIS and the FNNKD are shown in Figure 8.1. Figure 8.1 indicates that the predicted RMSEs of the FNNKD are smaller than those of ANFIS. Therefore, the FNNKD is a useful fuzzy neural network.

Table 8.2: Training Speeds (in Seconds) for 500 Epochs.

Model	Case 2	Case 3	Case 4	Case 5
ANFIS	535	280	147	61
FNNKD	45	26	13	6

8.2 Chaotic Time Series Prediction

The chaotic Mackey-Glass (MG) differential delay equation [43] is defined by:

$$\dot{x}(t) = \frac{0.2x(t-\tau)}{1+x^{10}(t-\tau)} - 0.1x(t). \tag{8.3}$$

The chaotic time series $(0 \le t \le 2000)$ is generated by the MG equation (8.3) when $x(0) = 1.2$, $\tau = 17$ and $x(t) = 0$ for $t < 0$. For comparison, we use the same 500 training data and 500 checking data used in [43], i.e, we use the following prediction format:

$$[x(t-18), x(t-12), x(t-6), x(t); x(t+6)], \tag{8.4}$$

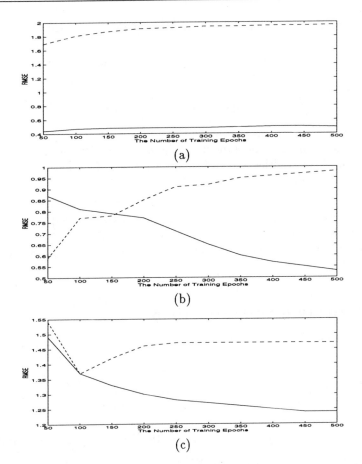

Figure 8.1. RMSEs (solid line) Generated by The FNNKD and RMSEs (dashed line) Generated by The ANFIS for (a) 27 Training Data and 8 Checking Data, (b) 64 Training Data and 27 Checking Data, (c) 125 Training Data and 64 Checking Data.

where $t = 118$ to 1117. An FNNKD for predicting the chaotic time series has 16 fuzzy rules such that

$$IF\ X_1\ is\ A_1^k\ and\ X_2\ is\ A_2^k\ and\ X_3\ is\ A_2^k\ and\ X_4\ is\ A_2^k \quad THEN\ Y\ is\ B^k,$$

where X_1, X_2, X_3 and X_4 are input fuzzy linguistic variables to represent $x(t-18)$, $x(t-12)$, $x(t-6)$ and $x(t)$, respectively, and Y is an output fuzzy linguistic variable to represent $x(t+6)$. A_i^k and B^k are defined by

$$\mu_{A_i^k}(x_i) = exp[-(\frac{x_i - a_i^k}{\sigma_i^k})^2], \tag{8.5}$$

$$\mu_{B^k}(y) = exp[-(\frac{y - b^k}{\eta^k})^2], \tag{8.6}$$

where a_i^k and b^k are centers of membership functions of x_i and y, respectively, and σ_i^k and η^k are widths of membership functions of x_i and y, respectively, for $i = 1, 2, 3, 4$ and $k = 1, 2, .., 16$.

In order to effectively analyze the performance of a FNNKD, in this section we discuss several aspects such as (1) comparison between Wang's fuzzy system and a FNNKD, (2) the effectiveness of the HGLA given in Section 4.3, (3) analysis of compensatory degrees, and (4) performance of an ANFIS, a FNNKD and the other systems.

8.2.1 Wang's Fuzzy System and a FNNKD

For comparison, we choose $\gamma^k = 0$, and then get a FNNKD described by:

$$f(x_1, ..., x_4) = \frac{\sum_{k=1}^{16}(b^k + \frac{\eta^k}{4}\sum_{i=1}^{4}\frac{w_i^k(x_i - a_i^k)}{\sigma_i^k})[\prod_{i=1}^{4}\mu_{A_i^k}(x_i^k)]}{\sum_{k=1}^{16}[\prod_{i=1}^{4}\mu_{A_i^k}(x_i^k)]}, \tag{8.7}$$

where x_1, x_2, x_3 and x_4 represent $x(t-18)$, $x(t-12)$, $x(t-6)$ and $x(t)$, respectively.

Wang's fuzzy system is a typical fuzzy system described by

$$g(x_1, ..., x_4) = \frac{\sum_{k=1}^{16} b^k [\prod_{i=1}^{4}\mu_{A_i^k}(x_i^k)]}{\sum_{k=1}^{16}[\prod_{i=1}^{4}\mu_{A_i^k}(x_i^k)]}. \tag{8.8}$$

After 500 training epochs for the 500 training data, the training RMSEs of the FNNKD and Wang's fuzzy system are 0.0036 and 0.0053, respectively. In addition, the checking RMSEs of the FNNKD and Wang's fuzzy system are 0.0036 and 0.0090, respectively. In addition, typical simulations shown in Figure 8.2 can also indicate that a FNNKD is better than Wang's fuzzy system in the related cases.

Sec. 8.2. Chaotic Time Series Prediction

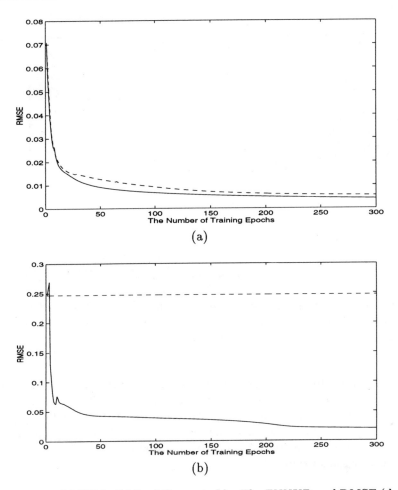

Figure 8.2. RMSE (solid line) Generated by The FNNKD and RMSE (dashed line) Generated by Wang's Fuzzy System When Both Use (a) Heuristic Parameters, (b) Random Initial Parameters.

8.2.2 Effectiveness of the HGLA

Because crisp weights in conventional neural networks have no apparent physical meaning, it is very hard to heuristically initialize them to prevent a neural network from reaching poor local minima. Importantly, all parameters in a FNNKD have physical meaning, therefore we can design the heuristic-

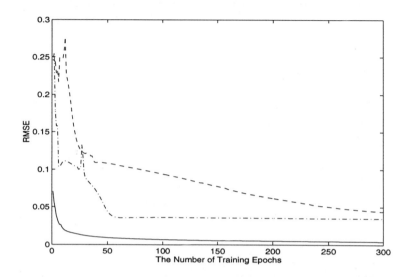

Figure 8.3. RMSEs (dashed line and dot-dashed line) Generated by The HGLA with Random Initial Parameters and RMSE (solid line) Generated by The HGLA with Heuristic Parameters.

knowledge-based HGLA to speed up convergence of the learning algorithm and find better solutions. To examine the efficiency of the HGLA, we compare a random-initialization-based learning algorithm with the HGLA by using the 500 training data. Typical simulation results shown in Figure 8.3 indicate that the learning speed of the HGLA is higher than that of a random-initialization-based learning algorithm.

8.2.3 Analysis of Compensatory Degrees γ^k

There are two ways to use compensatory degrees γ^k. The first way is to initialize γ^k and don't use the HGLA to train them. The simulation results shown in Figure 8.4(a) indicate that HGLAs with different initial γ^k have different learning speeds, and setting $\gamma^k = 0$ may be better because it can reduce the amount of calculation for the HGLA. The second way is to initialize γ^k and then use the HGLA to train them. The simulation results shown in Figure 8.4(b) demonstrate that the HGLA is able to quickly train a FNNKD under different initial γ^k.

Sec. 8.2. Chaotic Time Series Prediction

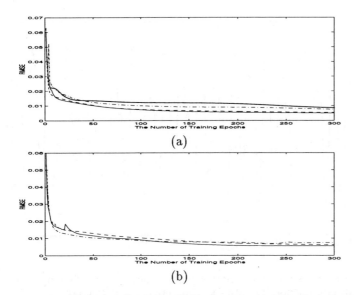

Figure 8.4. (a) RMSE (solid line) for $\gamma^k = 0$, RMSE (dashed line) for $\gamma^k = 0.1$, RMSE (dot-dashed line) for $\gamma^k = 0.2$ and RMSE (dot line) for $\gamma^k = 0.3$. (b) RMSE (solid line) for Initial $\gamma^k = 0.3$, RMSE (dashed line) for Initial $\gamma^k = 0.4$ and RMSE (dot-dashed line) for Initial $\gamma^k = 0.5$.

8.2.4 Performance of Various Approaches

The NDEI (Non-Dimensional Error Index) is defined as the RMSE divided by the standard deviation of the target series [43]. The results shown in Figure 8.5 and Table 8.3 indicate that the FNNKD is effective because it has not only low NDEI but also very short training time.

Table 8.3: Various Methods for Predicting Chaotic Time Series.

Method	Training Cases	NDEI
FNNKD	500	0.016
Wang's Fuzzy System	500	0.040
ANFIS	500	0.007
AR Model	500	0.19
Cascaded-Correlation NN	500	0.06
Back-Prop NN	500	0.02
Sixth-order Polynomial	500	0.04
Linear Predictive Method	2000	0.55

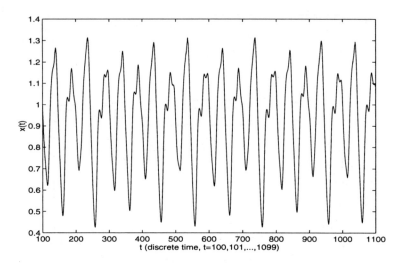

Figure 8.5. Desired Time Series (solid line) and Predicted Time Series (dashed line) after 500 Training Epochs (Using a FNNKD).

8.3 Box and Jenkins's Gas Furnace Model Identification

Box and Jenkins's gas furnace data is a frequently used benchmark for checking performance of fuzzy logic identification algorithms and other non-fuzzy ones [6,11,57,64,81,87,91,93,96,104,105,107,116]. The data set has 296 pairs of inputs (the gas flow rates) and outputs (the concentrations of CO_2). Since these fuzzy models used different numbers of inputs, these performance analyses are not reasonable and not effective to verify different fuzzy models. In order to compare our model (i.e., FNNKD) with the other ones objectively, we must do simulations under the same conditions (i.e., the same number of inputs and the same number of fuzzy rules (see Tables 8.4 and 8.5)). The simulation results are listed in Table 8.6.

Table 8.4: Fuzzy Models ($y(k)$: Desired Values, $\hat{y}(k)$: Predicted Values).

Fuzzy Model $\hat{y}(k)$	Mean Squared Error
$F(y(t-1), u(t-4))$	$\frac{1}{292}\sum_{k=5}^{296}[y(k)-\hat{y}(k)]^2$
$F(y(t-1), u(t-4), u(t-3))$ (in [77])	$\frac{1}{292}\sum_{k=5}^{296}[y(k)-\hat{y}(k)]^2$
$F(y(t-2), y(t-1), u(t-3))$ (in [6])	$\frac{1}{293}\sum_{k=4}^{296}[y(k)-\hat{y}(k)]^2$
$F(y(t-3), y(t-2), y(t-1), u(t-3), u(t-2), u(t-1))$	$\frac{1}{246}\sum_{k=51}^{296}[y(k)-\hat{y}(k)]^2$

Table 8.5: Conventional Models for The Gas Furnace Model Identification.

Model Name	Number of Inputs	Number of Rules	MSE
Box-Jenkins's model [11] (1976)	6	-	0.202
Linear model [93] (1993)	5	-	0.193
Saleem et al [87] (1987)	3 (4)	-	0.403
Saleem et al [87] (1987)	4	-	0.417

Table 8.6: Various Models for The Gas Furnace Model Identification.

Model Name	Number of Inputs	Number of Rules	MSE
TSK model [91] (1991)	2	2	0.359
Our model	2	2	0.149
Wang et al [105] (1996)	2	5	0.158
Our model	2	5	0.118
Kim et al [57] (1996)	2	11	0.108
Our model	2	11	0.095
Tong's model [96] (1980)	2	19	0.469
Our model	2	19	0.079
Pedrycz's model [81] (1984)	2	25	0.776
Xu's model [107] (1989)	2	25	0.328
Kupper's model [64] (1994)	2	25	0.166
Our model	2	25	0.072
Pedrycz's model [81] (1984)	2	49	0.478
Our model	2	49	0.064
Pedrycz's model [81] (1984)	2	81	0.320
Our model	2	81	0.061
Sugeno et al [93] (1993)	3	6	0.190
Our model	3	6	0.103
Bastian et al [6] (1996)	3	8	0.063
Our model	3	8	0.038
Sugeno-Tanaka's model [87] (1991)	6	2	0.068
Wang et al [104] (1995)	6	2	0.066
Zikidis et al [116] (1996)	6	2	0.064
Our model	6	2	0.058

A typical result using 5 fuzzy rules for $F(y(t-1), u(t-1))$ is shown in Figure 8.6. These simulation results in Table 8.6 and in Figure 8.6 have strongly indicated that (1) our model FNNKD is more effective compared with the other fuzzy models (see Table 8.6), (2) our model error decreases with the increasing of the number of rules (see Figure 8.7), (3) the HGLA is efficient and fast (see the small learning epochs in Table 8.6), and (4) our model (FNNKD) has the strong ability to approximate a complex nonlinear function. In addition, our model has set new records of lowest MSEs under the same conditions in Table 8.6. Therefore, the learning algorithm of the FNNKD is an efficient identification algorithm for the complex nonlinear system compared with the other 15 models from 1976 to 1996 in Tables 8.5 and 8.6.

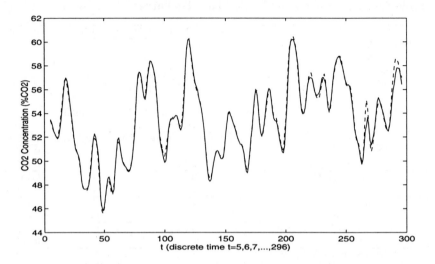

Figure 8.6. Desired Values (solid line) and Trained Values (dashed line) of CO_2 Concentration.

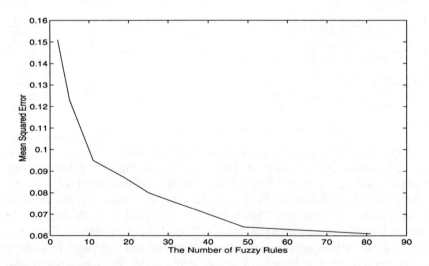

Figure 8.7. Relationship Between Mean Squared Error and The Number of Fuzzy Rules of Our Model for $y(t) = F(y(t-1), u(t-4))$.

Chapter 9

Fuzzy Moves in Fuzzy Games

In the four previous chapters, we have investigated various cases and benchmarks in order to illustrate our novel methods and models. In this chapter, we use the new normal fuzzy reasoning method and the effective hybrid system in making decisions for fuzzy moves. As we know, game theory is a very useful tool in economy, international relations, military strategy, and other decision-making areas. For instance, the Nobel prize on Economic Sciences in 1994 was awarded jointly to three pioneers of game theory. Obviously, using neuro-fuzzy systems in fuzzy games is an important tool with potential applications in various decision-making fields.

9.1 Introduction

The theory of fuzzy moves (TFM) [52,114] can be used to select relatively rational moves in fuzzy games in order to make a player's own payoff as high as possible as well as make his opponent's payoff as low as possible. Generally, the theory of fuzzy moves is a hybrid theory merging relevant principles and methods of the theory of moves (TOM) [12-14] and fuzzy sets theory. Some methods of making fuzzy moves were based on the conventional fuzzy reasoning methodology. For example, a conventional fuzzy reasoning method was used to perform a nonlinear transformation mapping from a local game to a global game [102], and neural networks were applied to learn such a nonlinear mapping from sample data [52]. We use a more reasonable fuzzy reasoning methodology to perform nonlinear game transformations. Importantly, the correctness and efficiency of the conventional fuzzy reasoning methodology can directly affect those of the methods of making fuzzy moves. If the conventional fuzzy reasoning methodology were perfect, then the methods of making fuzzy moves would be optimal. Unfortunately, since the conventional fuzzy reasoning methodology is not useful and not rational under some circumstances, we have to develop a more reasonable fuzzy reasoning methodology for making better fuzzy moves. Therefore, we use new primary fuzzy sets with higher data granularity than commonly used fuzzy sets to construct primary fuzzy rule bases which can contain more heuristic knowledge and useful information. Then we develop a new fuzzy reasoning methodology for fuzzy moves.

9.2 Fuzzy Moves

Here, we only consider a 2-player-2-strategy game as follows,

$$G_{2\times 2} = \begin{bmatrix} (a_{11}, b_{11}) & (a_{12}, b_{12}) \\ (a_{21}, b_{21}) & (a_{22}, b_{22}) \end{bmatrix}, \qquad (9.1)$$

where a_{ij} and b_{ij} for $i, j = 1, 2$ are payoffs of player A (i.e., row player) and player B (i.e., column player) respectively. Suppose $a_{ij} \in [0, 1]$ and $b_{ij} \in [0, 1]$.

Definition 9.1: Player A's absolutely changing quantity of payoffs between (a_{ij}, b_{ij}) and (a_{kl}, b_{kl}) is defined by

$$P_a(ij, kl) = a_{kl} - a_{ij}, \qquad (9.2)$$

player B's absolutely changing quantity of payoffs between (a_{ij}, b_{ij}) and (a_{kl}, b_{kl}) is defined by

$$P_b(ij, kl) = b_{kl} - b_{ij}. \qquad (9.3)$$

Definition 9.2: The local goal of each player in a game is to reach an outcome that is as advantageous to his own side as possible only based on his own payoffs. A game with only local goals is called a local game.

Definition 9.3: The global goal of each player in a game consists of two subgoals which are (1) maximizing his own payoff with the corresponding weight and (2) minimizing his opponent's payoff with the corresponding weight. In a global game with global goals, a player's decision is based on the importance, or weight, he assigns to each of the subgoals.

For simplicity, let P_a and P_b denote $P_a(ij, kl)$ and $P_b(ij, kl)$ respectively. We define triangular membership functions of P_a and P_b as follows,

$$\mu_{A_i}(P_a) = \begin{cases} 0 & \text{for } -1 \leq P_a \leq (\alpha_i - \delta_i) \\ 1 + \frac{P_a}{\delta_i} - \frac{\alpha_i}{\delta_i} & \text{for } (\alpha_i - \delta_i) < P_a \leq \alpha_i \\ 1 - \frac{P_a}{\delta_i} + \frac{\alpha_i}{\delta_i} & \text{for } a_{ij} < P_a < (\alpha_i + \delta_i) \\ 0 & \text{for } (\alpha_i + \delta_i) \leq P_a \leq 1. \end{cases} \qquad (9.4)$$

Sec. 9.2. Fuzzy Moves

$$\mu_{B_j}(P_b) = \begin{cases} 0 & \text{for } -1 \le P_b \le (\beta_j - \eta_j) \\ 1 + \frac{P_b}{\eta_j} - \frac{\beta_j}{\eta_j} & \text{for } (\beta_j - \eta_j) < P_b \le \beta_j \\ 1 - \frac{P_b}{\eta_j} + \frac{\beta_j}{\eta_j} & \text{for } \beta_j < P_b \le (\beta_j + \eta_j) \\ 0 & \text{for } (\beta_j + \eta_j) < P_b \le 1 \end{cases} \quad (9.5)$$

where α_i and β_j are centers of triangular fuzzy sets A_i for player A and B_j for player B, respectively, and δ_i and η_j are widths of triangular fuzzy sets A_i for player A and B_j for player B, respectively, for $i = 1, 2, ..., n$ and $j = 1, 2, ..., m$.

According to Definitions 9.2 and 9.3, the triangular membership functions of player A's global payoff P_c is defined as follows,

$$\mu_{C_k}(P_c) = \begin{cases} 0 & \text{for } -1 \le P_c \le (c_k - \sigma_k) \\ 1 + \frac{P_c}{\sigma_k} - \frac{c_k}{\sigma_k} & \text{for } (c_k - \sigma_k) < P_c \le c_k \\ 1 - \frac{P_c}{\sigma_k} + \frac{c_k}{\sigma_k} & \text{for } c_k < P_c < (c_k + \sigma_k) \\ 0 & \text{for } (c_k + \sigma_k) \le P_c \le 1 \end{cases} \quad (9.6)$$

where c_k are centers of triangular fuzzy sets C_k for player A and σ_k are widths of triangular fuzzy sets C_k for player A for $k = 1, 2, ..., N$. Generally, $c_k = f(\alpha_i, \beta_j)$.

Similarly, the triangular membership functions of player B's global payoff P_d is defined as follows,

$$\mu_{D_l}(P_d) = \begin{cases} 0 & \text{for } -1 \le P_d \le (d_l - \lambda_l) \\ 1 + \frac{P_d}{\lambda_l} - \frac{d_l}{\lambda_l} & \text{for } (d_l - \lambda_l) < P_d \le d_l \\ 1 - \frac{P_d}{\lambda_l} + \frac{d_l}{\lambda_l} & \text{for } d_l < P_d < (d_l + \lambda_l) \\ 0 & \text{for } (d_l + \lambda_l) \le P_d \le 1 \end{cases} \quad (9.7)$$

where d_l are centers of triangular fuzzy sets D_l for player B and λ_l are widths of triangular fuzzy sets D_l for player B for $l = 1, 2, ..., M$. Generally, $d_k = g(\beta_j, \alpha_i)$.

Based on the above defined various fuzzy sets for player A and B, we can easily get fuzzy rules such that:

(1) For player A

IF P_a is A_i and P_b is B_j THEN P_c is C_k;

(2) For player B

IF P_a is A_i and P_b is B_j THEN P_d is D_l.

9.3 Normal Fuzzy Reasoning for Fuzzy Moves

To understand the normal fuzzy reasoning, we only consider how to discover heuristic knowledge in a given n-by-n fuzzy rule base so as to construct a 2-input-1-output fuzzy system. By assuming $i, j = 1, 2, ..., n, k = 0, 1, 2, ..., n^2-1$, a 2-input-1-output fuzzy system $f_a(P_a, P_b)$ for player A with normal fuzzy reasoning is given below,

$$f_a(P_a, P_b) = \frac{\sum_{k=0}^{n^2-1} \phi_k [\mu_{A_k}(P_a)\mu_{B_k}(P_b)]^{1-\frac{\gamma_k}{2}}}{\sum_{k=0}^{n^2-1} [\mu_{A_k}(P_a)\mu_{B_k}(P_b)]^{1-\frac{\gamma_k}{2}}}, \qquad (9.8)$$

where

$$\phi_k = c_k + \frac{\sigma_k}{2}\left[\frac{w_k^a(P_a - \alpha_k)}{\delta_k} + \frac{w_k^b(P_b - \beta_k)}{\eta_k}\right].$$

Importantly, heuristic parameters w_k^a and w_k^b are defined below,

$$w_k^a = \begin{cases} w_k^{aleft} & \text{for } x_1 \leq \alpha_k \\ w_k^{aright} & \text{for } x_1 > \alpha_k, \end{cases} \qquad (9.9)$$

$$w_k^b = \begin{cases} w_k^{bleft} & \text{for } x_2 \leq \beta_k \\ w_k^{bright} & \text{for } x_2 > \beta_k, \end{cases} \qquad (9.10)$$

The KRA is used to get heuristic parameters w_k^{aleft}, w_k^{aright}, w_k^{bleft} and w_k^{bright} from a fuzzy rule base. For clarity, the KRA is given below:

Begin
$\quad N = n^2;$
$\quad$ for $(k = 0; k < n; k + +)$
$\quad\quad$ {if $(c_k < c_{k+n})$
$\quad\quad\quad$ then $\{w_k^{aleft} = 1; w_k^{aright} = 1;\}$
$\quad\quad$ else {if $(c_k > c_{k+n})$
$\quad\quad\quad$ then $w_k^{aleft} = -1; w_k^{aright} = -1;\}$
$\quad\quad\quad$ else $w_k^{aleft} = 0; w_k^{aright} = 0;\}$
$\quad\quad$ }
$\quad$ }
$\quad$ for $(k = n; k < (N - n); k + +)$
$\quad\quad$ {if $(c_k < c_{k-n})$
$\quad\quad\quad$ then $\{w_k^{aleft} = -1;\}$
$\quad\quad$ else {if $(c_k > c_{k-n})$

$$\text{then } \{w_k^{aleft} = 1;\}$$
$$\text{else } \{w_k^{aleft} = 0;\}$$
  }
  if $(c_k < c_{k-n})$
    then $\{w_k^{aright} = 1;\}$
    else {if $(c_k > c_{k-n})$
      then $\{w_k^{aright} = -1;\}$
      else $\{w_k^{aright} = 0;\}$
    }
  }
for $(k = (N-n); k < N; k++)$
  {if $(c_k < c_{k-n})$
    then $\{w_k^{aleft} = -1; w_k^{aright} = -1;\}$
    else {if $(c_k > c_{k-n})$
      then $\{w_k^{aleft} = 1; w_k^{aright} = 1;\}$
      else $\{w_k^{aleft} = 0; w_k^{aright} = 0;\}$
    }
  }
}
End.

According to the above methodology, we can also get a 2-input-1-output fuzzy system $f_b(P_a, P_b)$ for player A with normal fuzzy reasoning.

9.4 Applicability of Various Methods

At first, we use the new method of fuzzy moves to play a game, then analyze the efficiencies and rationale of both the new methodology of fuzzy moves and the old methodology of precise moves. Secondly, we examine the performance of conventional fuzzy reasoning methods and new normal fuzzy reasoning ones for making better fuzzy moves.

9.4.1 Prisoners' Dilemma

Here, we only consider a 2 × 2 game called prisoners' dilemma [83] as follows,

$$G_{2\times 2} = \begin{bmatrix} (0.45, 0.50) & (0.85, 0.35) \\ (0.25, 0.90) & (0.65, 0.75) \end{bmatrix}. \quad (9.11)$$

This prisoners' dilemma can also be described by the 2 × 2 game with 4 ordinal payoffs as follows,

$$\bar{G}_{2\times 2} = \begin{bmatrix} (2,2) & (4,1) \\ (1,4) & (3,3) \end{bmatrix}. \quad (9.12)$$

According to the standard game theory, (0.450, 0.550) (i.e., (2, 2)) is a Nash equilibrium because neither player A nor player B has an incentive to depart unilaterally from this outcome. According to the theory of moves [13], each starting state has its own outcome(s) based on clockwise and counter-clockwise progressions. For example, if (3, 3) is a starting state, then the outcome will also be (3, 3). Unfortunately, ordinal payoffs 1,2,3 and 4 are too precise to describe uncertain knowledge and quantitative information for making better fuzzy moves. In order to effectively play fuzzy games, we developed the theory of fuzzy moves merging the theory of moves and fuzzy sets theory. Now an interesting question is "Are fuzzy moves generated by the conventional fuzzy reasoning methodology worse than those generated by the new primary-fuzzy-sets-based fuzzy reasoning methodology?". For this problem, we show an example. Suppose: player A has two different global strategies denoted by GSA_1 and GSA_2 which are described in Tables 9.1 and 9.2, respectively, and player B has two different global strategies denoted by GSB_1 and GSB_2 which are described in Tables 9.3 and 9.4, respectively.

Table 9.1: 5×5 Fuzzy Rule Base with GSA_1 for Player A.

	$-\tilde{1}$	$-\tilde{0.5}$	$\tilde{0}$	$\tilde{0.5}$	$\tilde{1}$
$-\tilde{1}$	$\tilde{0.5}$	$\tilde{0.125}$	$-\tilde{0.25}$	$-\tilde{0.625}$	$-\tilde{1}$
$-\tilde{0.5}$	$\tilde{0.625}$	$\tilde{0.25}$	$-\tilde{0.125}$	$-\tilde{0.5}$	$-\tilde{0.875}$
$\tilde{0}$	$\tilde{0.75}$	$\tilde{0.375}$	$\tilde{0}$	$-\tilde{0.375}$	$-\tilde{0.75}$
$\tilde{0.5}$	$\tilde{0.875}$	$\tilde{0.5}$	$\tilde{0.125}$	$-\tilde{0.25}$	$-\tilde{0.625}$
$\tilde{1}$	$\tilde{1}$	$\tilde{0.625}$	$\tilde{0.25}$	$-\tilde{0.125}$	$-\tilde{0.5}$

Table 9.2: 5×5 Fuzzy Rule Base with GSA_2 for Player A.

	$-\tilde{1}$	$-\tilde{0.5}$	$\tilde{0}$	$\tilde{0.5}$	$\tilde{1}$
$-\tilde{1}$	$-\tilde{0.5}$	$-\tilde{0.625}$	$-\tilde{0.75}$	$-\tilde{0.875}$	$-\tilde{1}$
$-\tilde{0.5}$	$-\tilde{0.125}$	$-\tilde{0.25}$	$-\tilde{0.375}$	$-\tilde{0.5}$	$-\tilde{0.625}$
$\tilde{0}$	$\tilde{0.25}$	$\tilde{0.125}$	$\tilde{0}$	$-\tilde{0.125}$	$-\tilde{0.25}$
$\tilde{0.5}$	$\tilde{0.625}$	$\tilde{0.5}$	$\tilde{0.375}$	$\tilde{0.25}$	$\tilde{0.125}$
$\tilde{1}$	$\tilde{1}$	$\tilde{0.875}$	$\tilde{0.75}$	$\tilde{0.625}$	$\tilde{0.5}$

Table 9.3: 5×5 Fuzzy Rule Base with GSB_1 for Player B.

	-1	-0.5	0	0.5	1
-1	0.5	0.625	0.75	0.875	1
-0.5	0.125	0.25	0.375	0.5	0.625
0	-0.25	-0.125	0	0.125	0.25
0.5	-0.625	-0.5	-0.375	-0.25	-0.125
1	-1	-0.875	-0.75	-0.625	-0.5

Table 9.4: 5×5 Fuzzy Rule Base with GSB_2 for Player B.

	-1	-0.5	0	0.5	1
-1	-0.5	-0.125	0.25	0.625	1
-0.5	-0.625	-0.25	0.125	0.5	0.875
0	-0.75	-0.375	0	0.375	0.75
0.5	-0.875	-0.5	-0.125	0.25	0.625
1	-1	-0.625	-0.25	0.125	0.5

Fuzzy sets A_k, B_k and C_k are defined by relevant fuzzy numbers in Table 9.1, where input triangular centers $a_k = \lfloor k/10 \rfloor - 1$ and $b_k = (k \bmod 5)/2 - 1$, input triangular widths $\alpha_{ij} = \beta_{ij} = 0.375$, $c_k = \alpha a_k - (1-\alpha)b_k$, and output triangular width $\sigma_k = 0.125$. By using the knowledge discovery algorithm in section 4, we can get all heuristic parameters $w_k^a = 1$ and $w_k^b = -1$ for $k = 0, 1, 2, ..., 24$. According to $f_a(P_a, P_a)$ defined in section 4, we get the normal fuzzy system $f_a(P_a, P_a)$ for GSA_1 as follows,

$$f_a(P_a, P_b) = \frac{\sum_{k=0}^{24} \psi_k [\mu_{A_k}(P_a) \mu_{B_k}(P_b)]^{1-\frac{\gamma_k}{2}}}{\sum_{k=0}^{24} [\mu_{A_k}(P_a) \mu_{B_k}(P_b)]^{1-\frac{\gamma_k}{2}}}, \qquad (9.13)$$

where $\gamma_k = 0$, and

$$\psi_k = c_k + \frac{\sigma_k}{2} \left[\frac{w_k^a (P_a - \alpha_k)}{\delta_k} + \frac{w_k^b (P_b - \beta_k)}{\eta_k} \right].$$

Similarly, we can get the other three normal fuzzy systems with GSA_2 for player A and GSB_1 and GSB_2 for player B. For simplicity, only $f_a(P_a, P_a)$ for GS_{a1} is completely described, as above.

If the initial state is $(0.50, 0.75)$, we have
(1) The Counterclockwise Progression

The results are shown in Table 9.5 after 3 steps.

Table 9.5: Counterclockwise Progression for the Initial State (0.65, 0.75).

	A	B	A	B	
A Starts	(0.65, 0.75)	(0.85, 0.35)	(0.45, 0.50)	(0.25, 0.90)	(0.65, 0.75)
Survivor	?	(0.45, 0.50)	(0.45, 0.50)	(0.25, 0.90)	

A crucial problem described by a question mark in Table 9.5 is "what is the last outcome?", i.e., "which one is better between (0.45, 0.50) and (0.65, 0.75) for player A?". Now $P_a = 0.55 - 0.45 = 0.2$ and $P_b = 0.75 - 0.50 = 0.25$, then $f_a(P_a, P_a) = -0.142$ for GSA_1. Because $-0.142 < 0$, "Don't Move", i.e., (0.45, 0.50) is the final outcome for the clockwise progression. Importantly, if player A changes his global strategy by choosing GSA_2 instead of GSA_1, then we can get $g_a(P_a, P_a) = 0.058$ for GSA_2. Because $0.058 > 0$, "Move", i.e., (0.65, 0.75) is the final outcome for the counterclockwise progression. Therefore, it is rational that the final outcome depends on what global strategies players choose. But if we use the theory of moves, we will get only the one final outcome (0.65, 0.75) (i.e., (3, 3) in [13]) even though player A chooses totally different global strategies such as GSA_1 and GSA_2.

(2) The Clockwise Progression

The results are shown in Table 9.6 after 3 steps.

Table 9.6: Clockwise Progression for the Initial State (0.65, 0.75).

	B	A	B	A	
B Starts	(0.65, 0.75)	(0.25, 0.90)	(0.45, 0.50)	(0.85, 0.35)	(0.65, 0.75)
Survivor	?	(0.45, 0.50)	(0.45, 0.50)	(0.85, 0.35)	

Similarly, we get (1) the final outcome is (0.45, 0.50) for GSB_1; (2) the final outcome is (0.65, 0.75) for GSB_2. But we will get the only one final outcome (0.65, 0.75) (i.e., (3, 3) in [13]) according to the theory of moves.

Finally, the final outcomes for different global strategies of player A and B are listed in Table 9.7. Obviously, the TFM with normal fuzzy reasoning can make better and more reasonable moves than the theory of moves with precise reasoning since different global strategies are taken into account by the TFM.

Table 9.7: Final Outcomes for the Initial State (0.65, 0.75).

	GSA_1, GSB_1	GSA_1, GSB_2	GSA_2, GSB_1	GSA_2, GSB_2
TOM	(0.65, 0.75)	(0.65, 0.75)	(0.65, 0.75)	(0.65, 0.75)
TFM	(0.45, 0.50)	(0.45, 0.50)/(0.65, 0.75)	(0.65, 0.75)/(0.45, 0.50)	(0.65, 0.75)

9.4.2 Applicability of Fuzzy Reasoning Methods

Now we consider how to make better fuzzy moves for the following 2 cases:
(1) $(0.450, 0.550) \longrightarrow (0.325, 0.675)$;
(2) $(0.550, 0.450) \longrightarrow (0.675, 0.325)$.

For the above problems, we still use the above normal fuzzy system (8.13) for GSA_1 to make fuzzy moves. For comparison, a typical conventional fuzzy system [103] is described below,

$$F_a(P_a, P_b) = \frac{\sum_{k=0}^{24} c_k \mu_{A_k}(P_a) \mu_{B_k}(P_b)}{\sum_{k=0}^{24} \mu_{A_k}(P_a) \mu_{B_k}(P_b)}. \tag{9.14}$$

Case 1: $(0.450, 0.550) \longrightarrow (0.325, 0.675)$

We have $P_a = 0.325 - 0.450 = -0.125$ and $P_b = 0.675 - 0.550 = 0.125$, then we get $f_a(P_a, P_a) = -0.042$ and $F_a(P_a, P_a) = 0$. Therefore, we get the same result, that is, "Don't Move" (i.e., Stay at $(0.450, 0.550)$) since $f_a(P_a, P_a) = -0.042 < 0$ and $F_a(P_a, P_a) = 0$. For this case, both methodologies reach the same conclusion.

Case 2: $(0.550, 0.450) \longrightarrow (0.675, 0.325)$ We have $P_a = 0.675 - 0.550 = 0.125$ and $P_b = 0.325 - 0.450 = -0.125$, then we get $f_a(P_a, P_a) = 0.042$ and $F_a(P_a, P_a) = 0$. Therefore, the result generated by the new methodology is "Move" (i.e., move to $(0.675, 0.325)$ from $(0.450, 0.550)$) since $f_a(P_a, P_a) = 0.042 > 0$. Unreasonably, the result generated by the old methodology is still "Don't Move" (i.e., stay at $(0.550, 0.450)$) since $F_a(P_a, P_a) = 0$. Because $P_a = 0.125$ and $P_b = -0.125$, player A can get the benefit if he moves to $(0.675, 0.325)$ from $(0.450, 0.550)$. Intuitively, "Move" is much better than "Don't Move". Even more unreasonably, the old methodology always makes a player stay at an original state for any value of P_a and P_b for $P_a, P_b \in [-0.125, 0.125]$ because of the result $F_a(P_a, P_a) = 0$. On the other hand, we can use the new methodology to get reasonable different values of $f_a(P_a, P_a)$ for $P_a, P_b \in [-0.125, 0.125]$. The values of $F_a(P_a, P_a)$ (dashed line) and $f_a(P_a, P_a)$ (solid lines) for $P_a, P_b \in [-0.125, 0.125]$ are typically shown in Figure 9.1. Therefore, the novel fuzzy reasoning methodology is more reasonable and more powerful to make fuzzy moves than the conventional one.

9.5 Efficient Precise Decision Systems for Fuzzy Moves

Here, we show how to get a precise decision system by simplifying a normal fuzzy system. For instance, we have a 5-by-5 Fuzzy Rule Base in Table 9.8 for

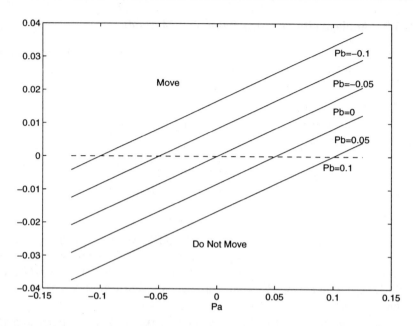

Figure 9.1. $F_a(P_a, P_a) = 0$ (Dashed Line) and $f_a(P_a, P_a)$ (Solid Lines) for $P_a \in [-0.125, 0.125]$ and $P_b = -0.1, -0.05, 0, 0.05, 1$.

Player A for $\alpha = 0.75$.

Table 9.8: 5-by-5 Fuzzy Rule Base for Player A for $\alpha = 0.75$.

	$\tilde{-1}$	$\tilde{-0.5}$	$\tilde{0}$	$\tilde{0.5}$	$\tilde{1}$
$\tilde{-1}$	$\tilde{-0.5}$	$\tilde{-0.625}$	$\tilde{-0.75}$	$\tilde{-0.875}$	$\tilde{-1}$
$\tilde{-0.5}$	$\tilde{-0.125}$	$\tilde{-0.25}$	$\tilde{-0.375}$	$\tilde{-0.5}$	$\tilde{-0.625}$
$\tilde{0}$	$\tilde{0.25}$	$\tilde{0.125}$	$\tilde{0}$	$\tilde{-0.125}$	$\tilde{-0.25}$
$\tilde{0.5}$	$\tilde{0.625}$	$\tilde{0.5}$	$\tilde{0.375}$	$\tilde{0.25}$	$\tilde{0.125}$
$\tilde{1}$	$\tilde{1}$	$\tilde{0.875}$	$\tilde{0.75}$	$\tilde{0.625}$	$\tilde{0.5}$

Suppose that input triangular centers $a_k = \lfloor k/10 \rfloor - 1$ and $b_k = (k \bmod 5)/2 - 1$, input triangular widths $\alpha_{ij} = \beta_{ij} = 0.375$, $c_k = \alpha a_k - (1-\alpha) b_k$, output triangular width $\sigma_k = 0.125$. By using the knowledge discovery algorithm in section 4, we can get all heuristic parameters $w_k^a = 1$ and $w_k^b = -1$ for $k = 0, 1, 2, ..., 24$.

Sec. 9.5. Efficient Precise Decision Systems for Fuzzy Moves

Table 9.9: Simple Mapping Functions Equivalent to $f_a(P_a, P_b)$ (Note: M: Move, DM: Don't Move).

P_a	P_b	Move Conditions	$Final Decisions$
$[-1, -0.375]$	$[-1, 1]$		DM.
$(-0.375, -0.125)$	$[-1, -0.875]$	$64P_a - 8P_b + 11 > 0$?	Yes, M; No, DM.
$(-0.375, -0.125)$	$(-0.875, -0.625]$	$16P_a - 4P_b + 1 > 0$?	Yes, M; No, DM.
$(-0.375, -0.125)$	$(-0.625, -0.375]$	$64P_a - 8P_b + 9 > 0$?	Yes, M; No, DM.
$(-0.375, -0.125)$	$(-0.375, -0.125]$	$32P_a - 8P_b + 3 > 0$?	Yes, M; No, DM.
$(-0.375, -0.125]$	$(-0.125, 1]$		DM.
$[-0.125, 0]$	$[-1, -0.375)$		M.
$[-0.125, 0]$	$[-0.375, -0.125]$	$8P_a - 16P_b - 1 > 0$?	Yes, M; No, DM.
$[-0.125, 0]$	$(-0.125, 0)]$	$P_a - P_b > 0$?	Yes, M; No, DM.
$[-0.125, 0]$	$[0, 1]$		DM.
$(0, 0.125]$	$[-1, 0)$		M.
$(0, 0.125]$	$[0, 0.125]$	$P_a - P_b > 0$?	Yes, M; No, DM.
$(0, 0.125]$	$(0.125, 0.375)]$	$8P_a - 16P_b + 1 > 0$?	Yes, M; No, DM.
$(0, 0.125]$	$[0.375, 1]$		DM.
$(0.125, 0.375]$	$[-1, 0.125]$		M.
$(0.125, 0.375]$	$(0.125, 0.375]$	$32P_a - 8P_b - 3 > 0$?	Yes, M; No, DM.
$(0.125, 0.375]$	$(0.375, 0.625]$	$64P_a - 8P_b - 9 > 0$?	Yes, M; No, DM.
$(0.125, 0.375]$	$(0.625, 0.875]$	$16P_a - 4P_b - 1 > 0$?	Yes, M; No, DM.
$(0.125, 0.375]$	$(0.875, 1]$	$64P_a - 8P_b - 11 > 0$?	Yes, M; No, DM.
$[0.375, 1]$	$[-1, 1]$		M.

According to $f_a(P_a, P_a)$ defined in section 3, we get the normal fuzzy system $f_a(P_a, P_a)$ for player A with $\alpha = 0.75$ as follows,

$$f_a(P_a, P_b) = \frac{\sum_{k=0}^{24}[c_k + \frac{1}{6}(P_a - P_b - \alpha_k + \beta_k)][\mu_{A_k}(P_a)\mu_{B_k}(P_b)]}{\sum_{k=0}^{24}[\mu_{A_k}(P_a)\mu_{B_k}(P_b)]}. \quad (9.15)$$

Now we can make decisions such that
IF $f_a(P_a, P_b) > 0$ THEN Move;
IF $f_a(P_a, P_b) \leq 0$ THEN Don't Move.

After carefully examining 25 fuzzy rules in Table 9.8, we can discover that in many cases we can directly check if $f_a(P_a, P_b) > 0$ without calculating the complex function $f_a(P_a, P_b)$. For example, if $-1 \leq P_a \leq -0.375$, then Don't Move (since $f_a(P_a, P_b) \leq 0$). In order to quickly make a decision for fuzzy moves, we can finally get the simple mathematical mapping functions listed in Table 9.9 by simplifying the function $f_a(P_a, P_b)$ for different regions.

9.6 Typical Examples

Here, we only consider a 2 × 2 game as follows,

$$G_{2\times 2} = \begin{bmatrix} (0.450, 0.550) & (0.550, 0.450) \\ (0.325, 0.675) & (0.675, 0.325) \end{bmatrix}. \tag{9.16}$$

Now we consider how to make better fuzzy moves for the following 2 cases:
(1) $(0.450, 0.550) \longrightarrow (0.325, 0.675)$;
(2) $(0.550, 0.450) \longrightarrow (0.675, 0.325)$.

For comparison, a typical conventional fuzzy system [103] is described below,

$$F_a(P_a, P_b) = \frac{\sum_{k=0}^{24} c_k \mu_{A_k}(P_a) \mu_{B_k}(P_b)}{\sum_{k=0}^{24} \mu_{A_k}(P_a) \mu_{B_k}(P_b)}. \tag{9.17}$$

Case 1: $(0.450, 0.550) \longrightarrow (0.325, 0.675)$

We have $P_a = 0.325 - 0.450 = -0.125$ and $P_b = 0.675 - 0.550 = 0.125$, then we get $f_a(P_a, P_a) = -0.042$ and $F_a(P_a, P_a) = 0$. Therefore, we get the same result, that is, "Don't Move"(i.e., Stay at (0.450, 0.550)), since $f_a(P_a, P_a) = -0.042 < 0$ and $F_a(P_a, P_a) = 0$. It is not efficient to calculate the value of $f_a(P_a, P_a)$ since we can easily get the decision "Don't Move" directly based on row 11 in Table 9.9. For this case, both methodologies reach the same conclusion.

Case 2: $(0.550, 0.450) \longrightarrow (0.675, 0.325)$

We have $P_a = 0.675 - 0.550 = 0.125$ and $P_b = 0.325 - 0.450 = -0.125$, then we get $f_a(P_a, P_a) = 0.042$ and $F_a(P_a, P_a) = 0$. Obviously, we can quickly get the decision "Move" directly based on row 12 in Table 9.9. Unreasonably, the result generated by the old methodology is still "Don't Move" (i.e., Stay at (0.550, 0.450)), since $F_a(P_a, P_a) = 0$. Because $P_a = 0.125$ and $P_b = -0.125$, player A can get the benefit if he moves to (0.675, 0.325) from (0.450, 0.550). Intuitively, "Move" is much better than "Don't Move". The old methodology always makes a player stay at an original state for any values of P_a and P_b for $P_a, P_b \in [-0.125, 0.125]$ since the unreasonable result $F_a(P_a, P_a) = 0$. Therefore, the simplified precise decision system for fuzzy moves is not only more reasonable to make fuzzy moves than the conventional system (9.17) but also more efficient than the complex normal fuzzy system (9.15).

A new normal fuzzy system, a generalized framework of Takagi-Sugeno's fuzzy system, is able to make more reasonable fuzzy reasoning based directly on given fuzzy rules. Through simplifying a normal fuzzy system for fuzzy moves, we can get a corresponding precise decision system for fuzzy moves, and then we can quickly decide whether to "Move" or not by just calculating some very simple functions without using the conventional fuzzification, fuzzy inference and defuzzification. The simulations have indicated that (1) the simplified precise decision system can make a fuzzy move more quickly than a normal fuzzy one and (2) a normal fuzzy system can make better and more reasonable moves than the conventional fuzzy systems. Therefore, the normal fuzzy decision system for fuzzy moves can make efficient and reasonable moves in fuzzy games.

9.7 Fuzzy Moves in Prisoner's Dilemma

Historically, the earlier versions of the Prisoner's Dilemma (PD) appeared in the writings of Seneca (4 B.C.-65 A.D.), Hillel (fl. 30 B.C.-9 A.D.), Aristotle (384-322 B.C.), Plato (427?-347 B.C.), and Confucius (551-479 B.C.) - and it may not have been original with any of them [83]. Its modern version was formulated by Dresher and Flood from RAND Corporation in 1950. Later, RAND consultant Albert W. Tucker dubbed the game PD. Since then a lot of research on the PD has been done because it is a mathematical construct and mirrors many real world situations where a conflict of interest exists. Price wars between competing companies, the arms race and overpopulation problems are examples of PD-like circumstances.

The traditional PD is shown by the payoff matrix in Figure 9.2. The paradox presented by the PD is that, regardless of the opponent's choice, the move "Confess" is the best for a player seeking the highest payoff in a single play, but if the other reasons in the same way, both end up much worse than if mutual "Not Confess" had been chosen instead. If the game is repeated, the situation becomes much more complex, and questions like "Is there an optimal strategy to be adopted in the PD?", or "What is an effective approach to the PD?" arise. In a well-known work, Axelrod [3] describes a computational tournament of the repeated PD, where several participants proposed their strategies, which were confronted to each other. The results indicated that a strategy called "tit-for-tat (TFT)" was the overall winner. TFT starts cooperating, and then simply repeats the opponent's last move. TFT proved to be quite robust, even when playing against other much more intricate rules.

Recently Brams [12-14] has developed what he called Theory of Moves (TOM) to add a dynamic dimension to classical Game Theory (GT). TOM also focuses on interdependent strategic situations in which the outcome depends on

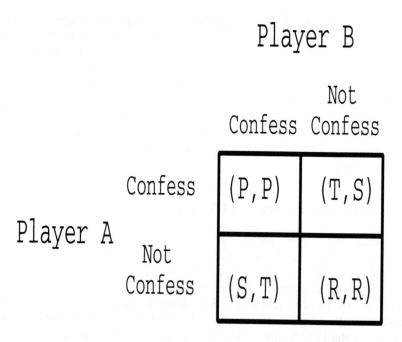

Figure 9.2. The Traditional PD (where, (a,b) =(payoff to Player A, payoff to Player B), $T > R > P > S$ and $R > (T+S)/2 > P$.)

the choices that all players make. It radically alters the rules of play, enabling players to look ahead–sometimes several steps–before making a move. There are two nonmyopic equilibria (NMEs) which are (P, P) and (R, R) in the PD according to TOM (see p.218, [13]).

However, many conventional methods assume that both players are egotists who play a *local* game, in the sense that a player only seeks the maximization of his (her) own gains and doesn't pays any attention to his (her) opponent's payoffs. If the contenders have different goals, the game's payoffs will be reevaluated and the outcomes may be quite different from those of the local game. Therefore, an original game should be transformed to a corresponding global game according to different global goals of players. In this sense, the outcome of a game depends not only on the original payoffs but also the global strategic goals of the players. In a global game, a player can evaluate each of the subgoals in a fuzzy manner, considering, for instance, the first subgoal very important and the second is less important. However, if a player didn't make

Sec. 9.7. Fuzzy Moves in Prisoner's Dilemma   129

a correct game transformation to express both his (her) own globally strategic goals and his (her) opponent's ones, he (she) would play a wrong or even totally wrong game. Actually, there are a lot of fuzzy data and uncertain information in dynamically changing environments. Therefore, it is necessary to develop a theory which is able to perform a nonlinear game transformation to effectively describe players' different globally strategic goals and finally make a reasonable decision for a complex game with uncertain information and fuzzy data.

9.7.1 Global Games and Global PDs

Basic Concepts

For clarity, we define some basic concepts for a 2-player-2-strategy game as follows,

Definition 9.4: An original 2-player-2-strategy game denoted by $g_{2\times 2}$ is defined by,

$$g_{2\times 2} = \begin{bmatrix} (x_{11}, y_{11}) & (x_{12}, y_{12}) \\ (x_{21}, y_{21}) & (x_{22}, y_{22}) \end{bmatrix}, \tag{9.18}$$

where x_{ij} and y_{ij} for $i,j = 1,2$ are raw payoffs of player A (i.e., row player) and player B (i.e., column player), respectively. These raw payoffs come directly from real situations.

Definition 9.5: A global 2-player-2-strategy game denoted by $G_{2\times 2}$ is defined by,

$$G_{2\times 2} = \begin{bmatrix} (a_{11}, b_{11}) & (a_{12}, b_{12}) \\ (a_{21}, b_{21}) & (a_{22}, b_{22}) \end{bmatrix}, \tag{9.19}$$

where a_{ij} and b_{ij} for $i,j = 1,2$ are global payoffs of player A (i.e., row player) and player B (i.e., column player), respectively. These global payoffs are generated by game transformation functions such that

$$a_{ij} = f_a(x_{ij}, y_{ij}), b_{ij} = f_b(x_{ij}, y_{ij}). \tag{9.20}$$

A game transformation function for a player represents a globally strategic goal of the player by taking both players' raw payoffs into account. Especially, a global game with $a_{ij} \in [0,1]$ and $b_{ij} \in [0,1]$ is called a normal global game.

Definition 9.5: A local 2-player-2-strategy game is a special global 2-player-2-strategy game with local game transformation functions such that

$$a_{ij} = f_a(x_{ij}), b_{ij} = f_b(y_{ij}). \tag{9.21}$$

A local game transformation function for a player represents a locally strategic goal of the player by taking only his (her) own raw payoff into account. Especially, a local game with $a_{ij} \in [0,1]$ and $b_{ij} \in [0,1]$ is called a normal local game.

According to the above definitions, before a player plays a game, he (she) has to subjectively map an original game to a global game so as to describe not only his (her) globally strategic goal but also his (her) opponent's one as objectively as possible. This natural procedure is called a game transformation such that

$$g_{2\times 2} \stackrel{f_a, f_b}{\Longrightarrow} G_{2\times 2}. \qquad (9.22)$$

Obviously, it is easier for a player to make his (her) own game transformation function to describe his (her) globally strategic goal, but much more difficult for the player to construct (or even guess) his (her) opponent's game transformation function so as to try to objectively describe his (her) opponent's globally strategic goal. However, if a player didn't correctly make his (her) own game transformation function or mistakenly guessed his (her) opponent's game transformation function, then he (she) would play a wrong or even totally wrong game. That is "if understanding both me and opponent, then getting a victory forever".

In a PD, if a player confesses and the other doesn't, the confessor goes free and the other gets a 100-year sentence; if both confess, both get an n-year sentence; and if both don't confess, they get an m-year sentence for $0 < m < 50 < n < 100$. For convenience, a game matrix is used to represent an original PD such that

$$pd = \begin{bmatrix} (n,n) & (0,100) \\ (100,0) & (m,m) \end{bmatrix}, \qquad (9.23)$$

where 0, 100, m and n are raw payoffs.

Now suppose $n = 60$ years and $m = 40$ years, then we have the original PD below,

$$pd = \begin{bmatrix} (60,60) & (0,100) \\ (100,0) & (40,40) \end{bmatrix}. \qquad (9.24)$$

This original PD is used to extensively analyze precise local and global PDs with precise game transformation functions in the following sub-sections. Assuming some *egotists*, *philanthropists*, *revengers* and *ascetics* are candidates to play the PD. Their different globally strategic goals with only one factor are typically listed in Table 9.10 (Note: some other more complex cases are not listed in Table 9.10).

Table 9.10: Players' Globally Strategic Goals with One Factor.

Players	Globally Strategic Goal
egotist	maximize his (her) own payoff
philanthropist	maximize his (her) opponent's payoff
ascetic	minimize his (her) own payoff
revenger	minimize his (her) opponent's payoff

Precise Local PDs

Case 1: an egotist and an egotist

In this case, meaningful precise game transformation functions are defined as
$$a_{ij} = \frac{100 - x_{ij}}{100}, b_{ij} = \frac{100 - y_{ij}}{100}. \quad (9.25)$$

After the game transformation, a commonly used PD in literature, actually a local game, is represented as

$$PD_{egoist-egoist} = \begin{bmatrix} (0.4, 0.4) & (1, 0) \\ (0, 1) & (0.6, 0.6) \end{bmatrix}. \quad (9.26)$$

The above local game transformation functions actually indicate that each of players only has a locally strategic goal since he (she) calculates a local payoff only based on his own raw payoff. Because such a local PD is commonly discussed in the classical GT and TOM, both GT and TOM have a "natural" assumption that is both players are all egotists each of whom only takes care of his (her) own payoffs and doesn't pay any attention to his (her) opponent's payoffs. Such a natural assumption really results in the dilemma in a PD since each player (egotist) tries to get his local payoff as much as possible. Therefore, an outcome is (0.4, 0.4) (i.e., Confess-Confess) according to the classical GT, and outcomes are (0.4,0.4)(i.e., Confess-Confess) and (0.6, 0.6) (i.e., Not Confess-Not Confess) according to TOM (see p.218,[13]). Strictly speaking, the above natural assumption is true only for some real situations but false for the other cases. Case 2 and 3 and the following global PDs in the next subsection will show incompleteness of the natural assumption.

Case 2: an egotist and an ascetic

In this game, game transformation functions are defined as
$$a_{ij} = \frac{100 - x_{ij}}{100}, b_{ij} = 1 - \frac{100 - y_{ij}}{100}, \quad (9.27)$$

then a new local PD is given by

$$PD_{egoist-ascetic} = \begin{bmatrix} (0.4, 0.6) & (1,1) \\ (0,0) & (0.6, 0.4) \end{bmatrix}. \qquad (9.28)$$

For this case, an outcome is $(1, 1)$ (i.e., Confess-Not Confess) according to GT, and an outcome is also $(1, 1)$ according to TOM. As a result, the egotist goes free and the ascetic gets a 100-year sentence.

Case 3: an ascetic and an ascetic

In this context, game transformation functions are defined as

$$a_{ij} = 1 - \frac{100 - x_{ij}}{100}, b_{ij} = 1 - \frac{100 - y_{ij}}{100}. \qquad (9.29)$$

After this game transformation, another local PD is described by

$$PD_{ascetic-ascetic} = \begin{bmatrix} (0.6, 0.6) & (0,1) \\ (1,0) & (0.4, 0.4) \end{bmatrix}. \qquad (9.30)$$

For such an ascetic-ascetic PD, an outcome is $(0.4, 0.4)$ (i.e., Not Confess-Not Confess) according to both GT and TOM. As a result, both ascetics are unhappy to get a 40-year sentence.

Obviously, game transformation functions are crucial to playing a game because they actually represent players' strategies by changing raw payoffs into meaningful global payoffs. More meaningful examples are discussed in the next sub-section.

Precise Global PDs

For completeness, we discuss all other 7 global PDs.

Case 1: an egotist and a philanthropist

For example, Bill is an egotist who only wants himself to be free, but his nice mother always wants her son to be free. For this special case, game transformation functions are

$$a_{ij} = \frac{100 - x_{ij}}{100}, b_{ij} = \frac{100 - x_{ij}}{100}, \qquad (9.31)$$

and then the resulting global PD is

$$PD_{egoist-philanthropist} = \begin{bmatrix} (0.4, 0.4) & (1,1) \\ (0,0) & (0.6, 0.6) \end{bmatrix}. \qquad (9.32)$$

Sec. 9.7. Fuzzy Moves in Prisoner's Dilemma

Now the reasonable outcome is $(1, 1)$ (i.e., Bill goes free, his mother gets a 100-year sentence). For such a real situation, if they still played the conventional PD, they would get a wrong outcome which is Confess-Confess.

Case 2: an egotist and a revenger

For this case, game transformation functions are

$$a_{ij} = \frac{100 - x_{ij}}{100}, b_{ij} = 1 - \frac{100 - x_{ij}}{100}, \qquad (9.33)$$

and then the global PD is

$$PD_{egoist-revenger} = \begin{bmatrix} (0.4, 0.6) & (1, 0) \\ (0, 1) & (0.6, 0.4) \end{bmatrix}. \qquad (9.34)$$

Finally, an interesting outcome is $(0.4, 0.6)$ (i.e., an egotist gets a 60-year sentence, so does a revenger).

Case 3: a philanthropist and a philanthropist

For this case, game transformation functions are

$$a_{ij} = \frac{100 - y_{ij}}{100}, b_{ij} = \frac{100 - x_{ij}}{100}, \qquad (9.35)$$

and then the global PD is

$$PD_{philanthropist-philanthropist} = \begin{bmatrix} (0.4, 0.4) & (0, 1) \\ (1, 0) & (0.6, 0.6) \end{bmatrix}. \qquad (9.36)$$

For such a real situation, the unhappy outcome is $(0.6, 0.6)$ (i.e., both philanthropists get a 40-year sentence).

Case 4: a philanthropist and a revenger

The game transformation functions

$$a_{ij} = \frac{100 - y_{ij}}{100}, b_{ij} = 1 - \frac{100 - x_{ij}}{100}, \qquad (9.37)$$

and then the global PD is

$$PD_{philanthropist-revenger} = \begin{bmatrix} (0.4, 0.6) & (0, 0) \\ (1, 1) & (0.6, 0.4) \end{bmatrix}. \qquad (9.38)$$

For such an interesting situation, both a revenger and a philanthropist are happy for the outcome $(1, 1)$ (i.e., the revenger goes free, but the philanthropist gets a 100-year sentence).

Case 5: an ascetic and a philanthropist

For this case, game transformation functions are

$$a_{ij} = 1 - \frac{100 - x_{ij}}{100}, b_{ij} = \frac{100 - x_{ij}}{100}, \quad (9.39)$$

and then the global PD is

$$PD_{ascetic-philanthropist} = \begin{bmatrix} (0.6, 0.4) & (0, 1) \\ (1, 0) & (0.4, 0.6) \end{bmatrix}. \quad (9.40)$$

For such an actual situation, both an ascetic and a philanthropist are unhappy for the outcome (0.4, 0.6) (i.e., both the ascetic and the philanthropist get a 40-year sentence).

Case 6: an ascetic and a revenger

For this case, game transformation functions are

$$a_{ij} = 1 - \frac{100 - x_{ij}}{100}, b_{ij} = 1 - \frac{100 - x_{ij}}{100}, \quad (9.41)$$

and then the global PD is

$$PD_{ascetic-revenger} = \begin{bmatrix} (0.6, 0.6) & (0, 0) \\ (1, 1) & (0.4, 0.4) \end{bmatrix}. \quad (9.42)$$

For such an unusual situation, the outcome is (1, 1) (i.e., the ascetic is very happy to get a 100-year sentence, and the revenger is also glad to go free).

Case 7: a revenger and a revenger

For this case, game transformation functions are

$$a_{ij} = 1 - \frac{100 - y_{ij}}{100}, b_{ij} = 1 - \frac{100 - x_{ij}}{100}, \quad (9.43)$$

and then the global PD is

$$PD_{revenger-revenger} = \begin{bmatrix} (0.6, 0.6) & (1, 0) \\ (0, 1) & (0.4, 0.4) \end{bmatrix}. \quad (9.44)$$

For this situation, the outcome is (0.6, 0.6) (i.e., both revengers are happy to get a 60-year sentence).

Sec. 9.7. Fuzzy Moves in Prisoner's Dilemma

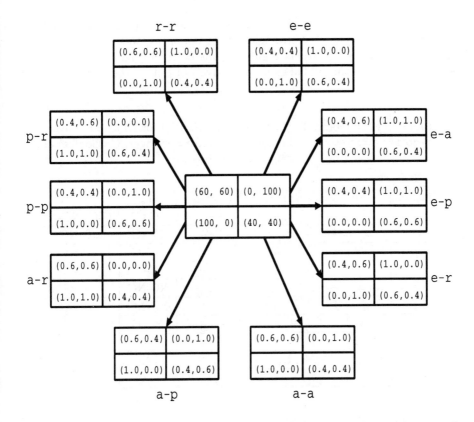

Figure 9.3. 10 Global PDs Transformed from the Original PD (e=egotist, a=ascetic, p=philanthropist and r=revenger).

Summary

For clarity, all results for the 10 local and global PDs are listed in Table 9.11, and 10 precise game transformations are depicted in Figure 9.3. Obviously, different players with different or same global strategic goals can make different decisions for an original PD since the original PD has been transformed to corresponding local or global PDs by game transformations. Consequently, if each player always took care of his (her) own local strategy and subjectively assumed that his (her) opponent was also an egotist using a similar local game transformation function, he (she) would play a wrong game with various players with different global strategic goals in complex situations. For completeness, a player has to fully consider all possible global game transformation functions by analyzing various global strategies of his opponent, then selects the most possible one, finally constructs a global game from an original game. At this step, since the resulting global game is relatively objective, the player can make a better move.

Table 9.11: Outcomes for Local and Global PDs for Different Players (e=egotist, a=ascetic, p=philanthropist and r=revenger, C=Confess and NC=Not Confess).

Players	Game Transformation Functions	GT	TOM
e-e	$a_{ij} = \frac{100-x_{ij}}{100}, b_{ij} = \frac{100-y_{ij}}{100}$	C-C	C-C, NC-NC
e-a	$a_{ij} = \frac{100-x_{ij}}{100}, b_{ij} = 1 - \frac{100-y_{ij}}{100}$	C-NC	C-NC
e-p	$a_{ij} = \frac{100-x_{ij}}{100}, b_{ij} = \frac{100-x_{ij}}{100}$	C-NC	C-NC
e-r	$a_{ij} = \frac{100-x_{ij}}{100}, b_{ij} = 1 - \frac{100-x_{ij}}{100}$	C-C	C-C
a-a	$a_{ij} = 1 - \frac{100-x_{ij}}{100}, b_{ij} = 1 - \frac{100-y_{ij}}{100}$	NC-NC	C-C
a-p	$a_{ij} = 1 - \frac{100-x_{ij}}{100}, b_{ij} = \frac{100-x_{ij}}{100}$	NC-NC	NC-NC
a-r	$a_{ij} = 1 - \frac{100-x_{ij}}{100}, b_{ij} = 1 - \frac{100-x_{ij}}{100}$	NC-C	NC-C
p-p	$a_{ij} = \frac{100-y_{ij}}{100}, b_{ij} = \frac{100-x_{ij}}{100}$	C-C	C-C
p-r	$a_{ij} = \frac{100-y_{ij}}{100}, b_{ij} = 1 - \frac{100-x_{ij}}{100}$	C-NC	C-NC
r-r	$a_{ij} = 1 - \frac{100-y_{ij}}{100}, b_{ij} = 1 - \frac{100-x_{ij}}{100}$	C-C	C-C

9.7.2 Theory of Fuzzy Moves

In the above section, precise linear game transformation functions are used to change an original PD into corresponding local and global PDs. Unfortunately, there exist a lot of nonlinear behaviors, uncertain information and ill-defined data in real games. Therefore, precise linear game transformation functions are hard to objectively describe nonlinear phenomena and uncertain data in complex games. In order to overcome the weakness of precise linear game transformation functions, we have developed TFM to deal with complex games with nonlinear fuzzy game transformation functions. TFM is a hybrid fuzzy game theory based on GT and TOM. The framework of TFM is represented in Figure 9.4.

For clarity, we introduce the principles of TFM layer by layer.
Layer 1: Fuzzy Game Transformation

Suppose an original game $g_{2 \times 2}$ with raw payoffs x_{ij} and y_{ij} for $i, j = 1, 2$

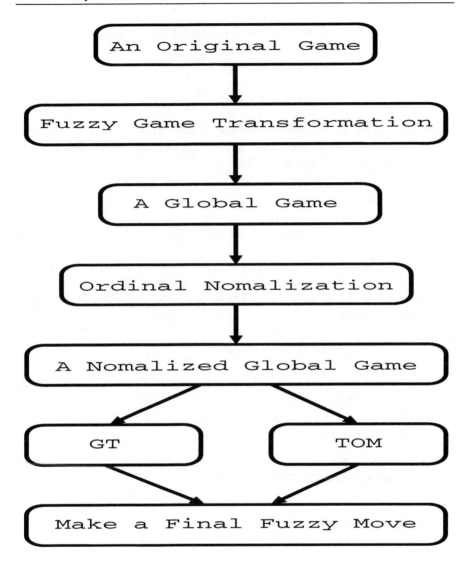

Figure 9.4. The Framework of Theory of Fuzzy Moves.

is given by

$$g_{2\times 2} = \begin{bmatrix} (x_{11}, y_{11}) & (x_{12}, y_{12}) \\ (x_{21}, y_{21}) & (x_{22}, y_{22}) \end{bmatrix}, \qquad (9.45)$$

then we define Gaussian membership functions of x_{ij} and y_{ij} as follows,

$$\mu_{A_k}(P_a) = exp[-(\frac{x_{ij} - \alpha_k}{\sigma_k})^2], \quad (9.46)$$

$$\mu_{B_k}(P_b) = exp[-(\frac{y_{ij} - \beta_k}{\delta_k})^2], \quad (9.47)$$

where α_k and σ_k are centers and widths of fuzzy sets A_k for player A, respectively, and β_k and δ_k are centers and widths of fuzzy sets B_k for player B, respectively, , $i, j = 1, 2$ and $k = 1, 2, ..., N$.

The Gaussian membership functions of player A's global payoff a_{ij} are defined as follows,

$$\mu_{C_k}(a_{ij}) = exp[-(\frac{a_{ij} - \theta_k}{\gamma_k})^2], \quad (9.48)$$

where θ_k and γ_k are centers and widths of fuzzy sets C_k for player A, respectively, $i, j = 1, 2$ and $k = 1, 2, ..., N$.

Similarly, the Gaussian membership functions of player B's global payoff b_{ij} are defined below,

$$\mu_{D_k}(b_{ij}) = exp[-(\frac{b_{ij} - \phi_k}{\eta_k})^2], \quad (9.49)$$

where ϕ_k and η_k are centers and widths of fuzzy sets D_k for player B, respectively, $i, j = 1, 2$ and $k = 1, 2, ..., N$.

Based on above defined various fuzzy sets for player A and B, we can easily make N fuzzy rules ($k = 1, 2, ..., N$) such that

(1) For player A

IF P_A is A_k and P_B is B_k THEN P_C is C_k;

(2) For player B

IF P_A is A_k and P_B is B_k THEN P_D is D_k.

Where P_A, P_B, P_C and P_D are fuzzy linguistic variables.

Now we need to find an efficient fuzzy reasoning method to make a better fuzzy game transformation. Based on the fuzzy system with modified center average defuzzifier (see page 22 in [103]), fuzzy game transformation functions are defined by

$$a_{ij} = \tilde{f}_a(x_{ij}, y_{ij}) = \frac{\sum_{k=1}^{N} c_k [\mu_{A_k}(x_{ij}) \mu_{B_k}(y_{ij}) / \gamma_k]}{\sum_{k=1}^{N} [\mu_{A_k}(x_{ij}) \mu_{B_k}(y_{ij}) / \gamma_k]}, \quad (9.50)$$

Sec. 9.7. Fuzzy Moves in Prisoner's Dilemma

$$b_{ij} = \tilde{f}_b(x_{ij}, y_{ij}) = \frac{\sum_{k=1}^{N} d_k[\mu_{A_k}(x_{ij})\mu_{B_k}(y_{ij})/\eta_k]}{\sum_{k=1}^{N}[\mu_{A_k}(x_{ij})\mu_{B_k}(y_{ij})/\eta_k]}, \quad (9.51)$$

where $i, j = 1, 2$ and $k = 1, 2, ..., N$.

After a fuzzy game transformation such that

$$g_{2\times 2} \stackrel{\tilde{f}_a, \tilde{f}_b}{\Longrightarrow} G_{2\times 2}, \quad (9.52)$$

the resulting global game is given by

$$G_{2\times 2} = \begin{bmatrix} (a_{11}, b_{11}) & (a_{12}, b_{12}) \\ (a_{21}, b_{21}) & (a_{22}, b_{22}) \end{bmatrix}. \quad (9.53)$$

Layer 2: Normalizing The Resulting Global Game

For convenience, global payoffs a_{ij} and b_{ij} for $i, j = 1, 2$ are normalized to *ordinal* payoffs $\bar{a}_{ij}$ and $\bar{b}_{ij}$ for $i, j = 1, 2$ by 4=best, 3=next best, 2=next worst and 1=worst (i.e., there are no ties), the normalized global game is finally given by

$$\bar{G}_{2\times 2} = \begin{bmatrix} (\bar{a}_{11}, \bar{b}_{11}) & (\bar{a}_{12}, \bar{b}_{12}) \\ (\bar{a}_{21}, \bar{b}_{21}) & (\bar{a}_{22}, \bar{b}_{22}) \end{bmatrix}, \quad (9.54)$$

where $\bar{a}_{ij}, \bar{b}_{ij} \in 1, 2, 3, 4$ for $i, j = 1, 2$.

Layer 3: Analyzing Outcome(s) by Using GT and TOM
Step 1: Use GT

By using the classical GT to analyze the resulting global game $\bar{G}_{2\times 2}$, we can get final outcome(s).

Step 2: Use TOM

TOM differs from the classical GT by assuming that players are in a particular state (initial state), from which either player can unilaterally switch its strategy, and thereby change the initial state into a new state. The other player can respond by unilaterally switching its strategy, hence moving the game to a new state. The alternating responses continue until the current player chooses not to switch its strategy. When this happens, the game terminates in a final state, which is the outcome of the game. In addition, a player will not move from an initial state if this move either (1) leads to a less preferred final outcome, or (2), brings the game back to the initial state. If it is rational for one player to move and the other player not to move from initial state, then the player who moves has precedence [12-14].

TOM assumes that the players can strictly rank the outcomes as follows: 4=best, 3=next best, 2=next worst, 1=worst. These payoffs are only ordinal:

they indicate only an ordering of outcomes from best to worst, but not the degree to which a player prefers one outcome over another.

By using TOM to analyzing the resulting global game $\bar{G}_{2\times 2}$, we can easily get final outcome(s) directly based on the outcome (4,4) for 21 no-conflict games and the given outcomes of 57 conflict games (see pp.215-219 in [13]).

Layer 4: Final Move Based on the above outcomes, a final fuzzy move is made by the player.

9.7.3 Fuzzy Moves in Global PDs

In Section 9.7.1, 10 global PDs are generated by precise game transformations with the globally strategic goals with only one factor described in Table 9.10, and 4 different players called *egotist, philanthropist, revenger* and *ascetic* join various PDs. Actually, an egotist-egotist PD is commonly studied in literature since it typically describes a most likely conflict situation. Therefore, , we discuss more complex egotist-egotist PDs with fuzzy game transformations with a two-factor globally strategic goal described in Table 9.12.

There are 4 players called Egotist A1, Egotist A2, Egotist B1 and Egotist B2 whose different globally strategic goals are listed in Table 9.12. These globally strategic goals of Egotist A1, Egotist A2, Egotist B1 and Egotist B2 are described by N fuzzy IF-THEN rules such that

$$IF\ P_A\ is\ A_k\ and\ P_B\ is\ B_k\ THEN\ P_C\ is\ C_k,$$

where P_A, P_B, P_C are fuzzy linguistic variables, A_k and B_k are Gaussian fuzzy numbers on $[0, 100]$, and C_k are Gaussian fuzzy numbers on $[0, 1]$ for $k = 1, 2, ..., N$. The detailed fuzzy rules used by Egotist A1, Egotist A2, Egotist B1 and Egotist B2 are represented in fuzzy rule bases in Tables 9.13, 9.14, 9.15 and 9.16, respectively. The Gaussian fuzzy numbers are defined by (29-32). Suppose $\sigma_k = \delta_k = 25$, $\gamma_k = \eta_k = 12.5$, and α_k, β_k, θ_k and ϕ_k are just the centers of corresponding Gaussian fuzzy numbers defined in fuzzy rule bases.

Table 9.12: Players' Globally Strategic Goals with Two Factors.

Player	Important Strategic Goal	Minor Strategic Goal
A1	maximize his own payoff	minimize his opponent's payoff
A2	minimize his opponent's payoff	maximize his own payoff
B1	maximize his own payoff	minimize his opponent's payoff
B2	minimize his opponent's payoff	maximize his own payoff

Sec. 9.7. Fuzzy Moves in Prisoner's Dilemma

Table 9.13: 5 × 5 Fuzzy Rule Base for Egotist A1.

	$\tilde{0}$	$\tilde{25}$	$\tilde{50}$	$\tilde{75}$	$\tilde{100}$
$\tilde{0}$	$0.\tilde{8}0$	$0.\tilde{8}5$	$0.\tilde{9}0$	$0.\tilde{9}5$	$\tilde{1}$
$\tilde{25}$	$0.\tilde{6}0$	$0.\tilde{6}5$	$0.\tilde{7}0$	$0.\tilde{7}5$	$0.\tilde{8}0$
$\tilde{50}$	$0.\tilde{4}0$	$0.\tilde{4}5$	$0.\tilde{5}0$	$0.\tilde{5}5$	$0.\tilde{6}0$
$\tilde{75}$	$0.\tilde{2}0$	$0.\tilde{2}5$	$0.\tilde{3}0$	$0.\tilde{3}5$	$0.\tilde{4}0$
$\tilde{100}$	$\tilde{0}$	$0.\tilde{0}5$	$0.\tilde{1}0$	$0.\tilde{1}5$	$0.\tilde{2}0$

Table 9.14: 5 × 5 Fuzzy Rule Base for Egotist A2.

	$\tilde{0}$	$\tilde{25}$	$\tilde{50}$	$\tilde{75}$	$\tilde{100}$
$\tilde{0}$	$0.\tilde{2}0$	$0.\tilde{4}0$	$0.\tilde{6}0$	$0.\tilde{8}0$	$\tilde{1}$
$\tilde{25}$	$0.\tilde{1}5$	$0.\tilde{3}5$	$0.\tilde{5}5$	$0.\tilde{7}5$	$0.\tilde{9}5$
$\tilde{50}$	$0.\tilde{1}0$	$0.\tilde{3}0$	$0.\tilde{5}0$	$0.\tilde{7}0$	$0.\tilde{9}0$
$\tilde{75}$	$0.\tilde{0}5$	$0.\tilde{2}5$	$0.\tilde{4}5$	$0.\tilde{6}5$	$0.\tilde{8}5$
$\tilde{100}$	$\tilde{0}$	$0.\tilde{2}0$	$0.\tilde{4}0$	$0.\tilde{6}0$	$0.\tilde{8}0$

Table 9.15: 5 × 5 Fuzzy Rule Base for Egotist B1.

	$\tilde{0}$	$\tilde{25}$	$\tilde{50}$	$\tilde{75}$	$\tilde{100}$
$\tilde{0}$	$0.\tilde{6}0$	$0.\tilde{7}0$	$0.\tilde{8}0$	$0.\tilde{9}0$	$\tilde{1}$
$\tilde{25}$	$0.\tilde{4}5$	$0.\tilde{5}5$	$0.\tilde{6}5$	$0.\tilde{7}5$	$0.\tilde{8}5$
$\tilde{50}$	$0.\tilde{3}0$	$0.\tilde{4}0$	$0.\tilde{5}0$	$0.\tilde{6}0$	$0.\tilde{7}0$
$\tilde{75}$	$0.\tilde{1}5$	$0.\tilde{2}5$	$0.\tilde{3}5$	$0.\tilde{4}5$	$0.\tilde{5}5$
$\tilde{100}$	$\tilde{0}$	$0.\tilde{1}0$	$0.\tilde{2}0$	$0.\tilde{3}0$	$0.\tilde{4}0$

Table 9.16: 5 × 5 Fuzzy Rule Base for Egotist B2.

	$\tilde{0}$	$\tilde{25}$	$\tilde{50}$	$\tilde{75}$	$\tilde{100}$
$\tilde{0}$	$0.\tilde{3}0$	$0.\tilde{4}75$	$0.\tilde{6}5$	$0.\tilde{8}25$	$\tilde{1}$
$\tilde{25}$	$0.\tilde{2}25$	$0.\tilde{4}0$	$0.\tilde{5}75$	$0.\tilde{7}5$	$0.\tilde{9}25$
$\tilde{50}$	$0.\tilde{1}5$	$0.\tilde{3}25$	$0.\tilde{5}0$	$0.\tilde{6}75$	$0.\tilde{8}5$
$\tilde{75}$	$0.\tilde{0}75$	$0.\tilde{2}5$	$0.\tilde{4}25$	$0.\tilde{6}0$	$0.\tilde{7}75$
$\tilde{100}$	$\tilde{0}$	$0.\tilde{1}75$	$0.\tilde{3}5$	$0.\tilde{5}25$	$0.\tilde{7}0$

For clarity, we follow the procedure depicted in Figure 9.3 to analyze 4 global egotist-egotist PDs for the original PD.

Step 1: Fuzzy Game Transformation

At first, we analyze the game between Egotist A1 and Egotist B1. By using the fuzzy game transformation functions Eq.(9.50) and (9.51) for Egotist A1 and Egotist B1 respectively, we can change the original PD into a corresponding global PD which is given by

$$PD_{EgoistA1-EgoistB1} = \begin{bmatrix} (0.44, 0.48) & (0.93, 0.07) \\ (0.07, 0.93) & (0.56, 0.52) \end{bmatrix}. \qquad (9.55)$$

Similarly, we can use the above method to analyze other PDs which are (Egotist A1 and Egotist B2), (Egotist A2 and Egotist B1) and (Egotist A2 and Egotist B2). Finally, all results are clearly described in Figure 9.4.

Step 2: Ordinal Normalization

By normalizing the above resulting global PD, we get the normalized global PD such that

$$NPD_{EgoistA1-EgoistB1} = \begin{bmatrix} (2,2) & (4,1) \\ (1,4) & (3,3) \end{bmatrix}. \qquad (9.56)$$

Similarly, we can get the other ordinal global PDs.

Step 3: Using GT and TOM

The outcomes of 4 normalized global PDs according to GT and TOM are represented in Figure 9.5, respectively. For example, according to GT, (2, 2) (i.e., Confess-Confess) is an outcome. Based on TOM, (2, 2) and (3, 3) are outcomes (see p.218, [13]).

Step 4: Final Fuzzy Moves

By analyzing outcomes generated by GT and TOM, final most likely outcomes for corresponding global PDs are given in Figure 9.5. For instance, for $PD_{EgoistA1-EgoistB1}$, (2, 2) should be a most likely outcome since it is an outcome generated by both GT and TOM (although (3, 3) is another outcome according to TOM).

In summary, it is relatively easy for players to express their globally strategic goals by using fuzzy IF-THEN rules such as IF (Player A's Payoff is *Very High*) and (Player B's Payoff is *Low*) THEN (Player A's Global Payoff is *High*). A fuzzy game transformation is an efficient way to describe complex nonlinear strategies of players by using commonly used linguistic words such as *Very High, Very Low* and *Not Very Low*. If a player mistakenly made a nonlinear game transformation, he (she) would play a wrong or even totally wrong

Sec. 9.7. Fuzzy Moves in Prisoner's Dilemma   143

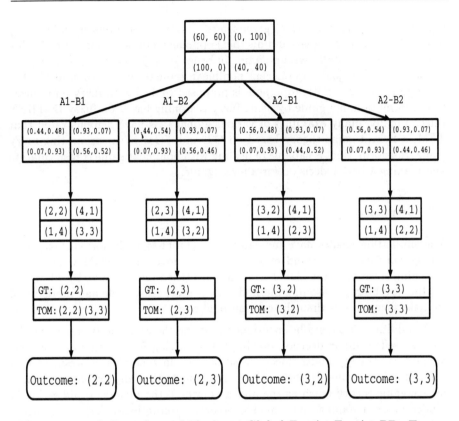

Figure 9.5. A Procedure of Playing 4 Global Egotist-Egotist PDs Transformed from the Original PD (A1=Egotist A1, A2=Egotist A2, B1=Egotist B1 and B2=Egotist B2).

game. However, how to effectively make a nonlinear game transformation such as fuzzy game transformation is still an open problem in the future.

9.7.4 Conclusions

Many conventional methods of GT and TOM often deal with *local* games in which a player only takes care of his (her) own payoffs but doesn't pays any attention to his (her) opponent's payoffs. The extensive analysis on PDs has indicated that these conventional methods are not effective for more complex games in which players have different globally strategic goals. Importantly, if a player didn't make a correct game transformation to express both his (her) own globally strategic goals and his (her) opponent's ones, he (she) would play a wrong or even totally wrong game. In order to objectively express players' globally strategic goals, an efficient nonlinear game transformation method, a fuzzy game transformation method, is proposed to deal with fuzzy and uncertain information in a game by using fuzzy reasoning based on fuzzy IF-THEN rules. Finally, many examples of various global PDs have demonstrated that TFM, a hybrid game theory based on GT and TOM by incorporating fuzzy game transformation, is capable of dealing with nonlinear game transformation and making a better decision for a fuzzy game.

9.8 Summary

The simulation results have indicated that (1) TFM with normal fuzzy reasoning can make better and more reasonable moves than TOM with precise reasoning since different global strategies are taken into account by TFM and (2) the novel fuzzy reasoning methodology is more reasonable and more useful to make fuzzy moves than the conventional one.

In addition, it should be mentioned that the theory of fuzzy moves can be widely used in many decision-making fields ranging from economic decisions to military strategy under uncertain situations. In general, if a president or a leader made a wrong decision nationally or internationally according to ineffective game theory (such as the above example), the mistake could seriously hinder or even damage nation-wide or even international defense, security, economy, or diplomacy. For example, President Jimmy Carter played a wrong game with Ayatollah Ruhollah Khomeini in the Iran hostage crisis in 1980, because he used the classical game theory. In the future, how to use a hybrid system based on fuzzy logic, neural networks, probability, genetic algorithms, etc. to play fuzzy games will be a very competitive research area.

Chapter 10

Genetic Neuro-fuzzy Pattern Recognition

Pattern recognition is an important application of both neural network techniques and fuzzy sets theory. Various crisp neural networks such as the Hopfield network [74], the Hamming net [74], the Carpenter/Grossberg ART [18], Martin-Pittman BP-based neural net [75] and the Fukushima Neocognitron [29,30] have been used in pattern recognition. On the other hand, fuzzy sets theory is able to deal with uncertainty and fuzziness in pattern recognition, therefore fuzzy logic is a useful tool for simulating the human brain's perception and decision. In recent years, fuzzy neural networks have been used in pattern recognition. Kosko proposed a Fuzzy Associate Memory (FAM) using fuzzy matrices to represent fuzzy mappings [60]. Kwan and Cai designed a fuzzy neural network to recognize the 26 English letters and the 10 Arabic numericals [65]. However, the Kwan-Cai Fuzzy Neural Network (FNN) has two disadvantages which are (1) the important parameter α is defined subjectively and (2) the learning algorithm may greatly increase the number of fuzzy neurons if many training patterns are not similar. To solve these problems, we have developed a genetic fuzzy neural network which is capable of using genetic algorithms to optimize the parameter α and reducing the number of fuzzy neurons. The simulations have indicated that the genetic fuzzy neural network can recognize various distorted patterns with high recognition rates.

10.1 Structure of a Genetic Fuzzy Neural Network

A Genetic Fuzzy Neural Network (GFNN) is a genetic-algorithms-based Kwan-Cai FNN, and consists of 4 layers (see Figure 10.1). Each pattern has $N_1 \times N_2$

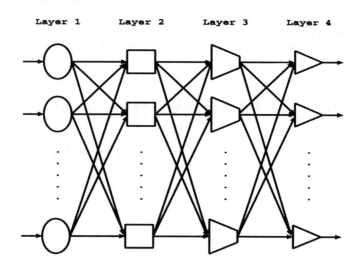

Figure 10.1. A Genetic Fuzzy Neural Network.

pixels and there are K training patterns. For clarity, the functions of fuzzy neurons in different layers are described layer by layer as follows:

Layer 1: Input Layer

$N_1 \times N_2$ input neurons in the first layer are oval nodes with simple normalization functions defined by

$$O_{ij}^{[1]} = \frac{x_{ij}^k}{max_{i=1}^{N_1}(max_{j=1}^{N_2}(max_{k=1}^{K}(x_{ij}^k)))}, \qquad (10.1)$$

where x_{ij}^k is the (i,j)th pixel value of the kth pattern for $i = 1, 2, ..., N_1$, $j = 1, 2, ..., N_2$ and $k = 1, 2, ..., K$.

Layer 2: Genetic Fuzzification Layer

$N_1 \times N_2$ Genetic fuzzification neurons in the second layer are square nodes with either the triangular fuzzification functions used by Kwan and Cai [65] or the new Gaussian fuzzification functions.

The triangular fuzzification functions are given by:

$$O_{pqm}^{[2]} = \begin{cases} 1 - \frac{2|\gamma_{pqm} - f_{pq}|}{\alpha} & \text{for } O_{pqm}^{[2]} \geq 0 \\ 0 & \text{for otherwise,} \end{cases} \qquad (10.2)$$

Sec. 10.2. Genetic-Algorithms-Based Self-Organizing Learning Algorithm   147

the Gaussian fuzzification functions are given by:

$$O^{[2]}_{pqm} = e^{-(\frac{\gamma_{pqm}-f_{pq}}{\alpha})^2}, \qquad (10.3)$$

where

$$f_{pq} = max_{i=1}^{N_1}(max_{j=1}^{N_2}(O^{[1]}_{ij}e^{-\beta^2[(p-i)^2+(q-j)^2]})), \qquad (10.4)$$

where γ_{pqm} and β are parameters to adjust centers and widths of fuzzification functions f_{pq}, respectively, and α is a parameter to change the shapes of functions $O^{[2]}_{pqm}$ for $p = 1, 2, ..., N_1$, $q = 1, 2, ..., N_2$ and $m = 1, 2, ..., M$.

Genetic algorithms are used to optimize α for K input patterns in order to reduce the number of neurons in the layers 2, 3 and 4.

Layer 3: Fuzzy Clustering Layer

M fuzzy neurons in the 3rd layer are trapezoidal nodes with a min function defined by

$$O^{[3]}_m = min^{N_1}_{p=1}(min^{N_2}_{q=1}(O^{[2]}_{pqm})), \qquad (10.5)$$

where $m = 1, 2, ..., M$. M will be determined by the learning algorithm.

Layer 4: Output Layer

M output neurons in the 4th layer are triangular nodes with the functions defined by

$$O^{[4]}_m = \begin{cases} 0 & \text{for } O^{[3]}_{pqm} < max^M_{m=1}(O^{[3]}_{pqm}) \\ 1 & \text{for } O^{[3]}_{pqm} = max^M_{m=1}(O^{[3]}_{pqm}), \end{cases} \qquad (10.6)$$

where $m = 1, 2, ..., M$.

10.2 Genetic-Algorithms-Based Self-Organizing Learning Algorithm

By adding genetic algorithms into the Kwan-Cai learning algorithm [65], we propose a genetic-algorithm-based self-organizing learning algorithm which is able to reduce M by adjusting α. The total number of fuzzy neurons in a GFNN is $(1+M)N_1N_2+2M$. N_1 and N_2 are fixed, therefore we can only try to reduce M to reduce the complexity of the GFNN. The genetic-algorithm-based self-organizing learning algorithm is given below,

Step 1: Choose values of $\alpha(\alpha \geq 0)$ and β for the second layer of the GFNN.

Step 2: $m = 0$ and $k = 1$.

Step 3: $m = m + 1$. Calculate:

$$\gamma_{pqm} = max_{i=1}^{N_1}(max_{j=1}^{N_2}(O_{ij}^{[1]}\theta)), \qquad (10.7)$$

where

$$\theta = e^{-\beta^2[(p-i)^2+(q-j)^2]}, \qquad (10.8)$$

for $p = 1, 2, ..., N_1$ and $q = 1, 2, ..., N_2$.

Step 4:

If $m < M_{max}$

Then goto *Step 5*.

Else use genetic algorithms to adjust α:

The fitness function is given by:

$$F = \sum_{k=1}^{K} \sum_{m=1}^{M_{max}} |O_m^k - y_m|, \qquad (10.9)$$

where O_m^k are the outputs of the 4th layer of the GFNN and y_m^k are the desired outputs for the kth training pattern.

Finally goto *Step 6*.

(Note: M_{max} is the maximum number of output neurons given by a user.)

Step 5: $k = k + 1$.

If $k > K$

Then goto *Step 6*.

Else calculate:

$$\sigma = 1 - max_{j=1}^{m}(O_{jk}^{[3]}), \qquad (10.10)$$

where $O_{jk}^{[3]}$ is the output of the jth neuron in the 3rd layer for the kth training pattern.

If $\sigma \leq T_f$

Then goto *Step 4*.

Else goto *Step 3*.

(Note: T_f is the fault tolerance of the GFNN.)

Step 6: End.

Figure 10.2. LA and SM distorted patterns (LA: larger, SM: smaller).

Figure 10.3. TT and SQ distorted patterns (TT: taller and thinner, SQ: squished).

10.3 Simulations

We use 10 kinds of the distorted 26 English letters and 10 Arabic numerals to verify the new GFNN. For example, 10 kinds of distorted patterns for "H" and "4" are shown in Figures 10.2, 10.3, 10.4, 10.5 and 10.6. In the simulations, $N_1 = N_2 = 16$, $M_{max} = 36$, $K = 72$, $\beta = 0.3$, the initial $\alpha = 2.0$. We used 36 LA & 36 SM, 36 TT & 36 SQ, and 36 SH & 36 DC distorted patterns to train a GFNN, respectively, then used 360 distorted patterns to check recognition rates. In order to compare the GFNN with the Kwan-Cai FNN, we adopted the triangular fuzzification function Eq. (10.2) in the simulations. By using genetic algorithms, we finally get $\alpha = 1.684$ for the trained GFNN.

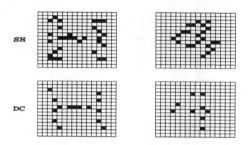

Figure 10.4. SH and DC distorted patterns (SH: shaking, DC: disconnected).

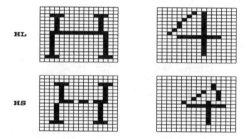

Figure 10.5. HL and HS distorted patterns (HL: half-part larger, HS: half-part shifted).

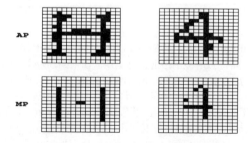

Figure 10.6. AP and MP distorted patterns (AP: added small parts, MP: missed small parts).

The simulation results are given in Tables 10.1, 10.2 and 10.3. As a result, the new GFNN is better than the Kwan-Cai FNN because the GFNN can optimize α by using genetic algorithms.

Table 10.1: *Comparison between the Kwan-Cai FNN's Recognition Rate and our GFNN's Recognition Rate for $T_f = 0.52$.*

Training Patterns	36 LAs and 36 SMs
Kwan-Cai FNN	87.78%
Our GFNN	94.44%

Table 10.2: *Comparison between the Kwan-Cai FNN's Recognition Rate and our GFNN's Recognition Rate for $T_f = 0.48$.*

Training Patterns	36 TTs and 36 SQs
Kwan-Cai FNN	87.78%
Our GFNN	93.05%

Table 10.3: *Comparison between the Kwan-Cai FNN's Recognition Rate and our GFNN's Recognition Rate for $T_f = 0.515$.*

Training Patterns	36 SHs and 36 DCs
Kwan-Cai FNN	87.78%
Our GFNN	89.17%

10.4 Conclusions

By applying genetic algorithms to the Kwan-Cai fuzzy neural network, we have designed a more adaptive genetic fuzzy neural network for pattern recognition. A genetic-algorithms-based self-organizing learning algorithm is capable of reducing the total number of fuzzy neurons and increasing recognition rates for a fixed number of output neurons. The simulations have indicated that the genetic fuzzy neural network is effective for recognizing various distorted patterns with good recognition rates.

Chapter 11

Constructive Approach to Modeling Fuzzy Systems

11.1 Introduction

Two important issues for designing a fuzzy logic system are (1) how to make fuzzy reasoning physically and effectively and (2) how to model a fuzzy system for both given data and a required error. For the first issue, many fuzzy reasoning methods applied various fuzzy logical implications to try to make reasonable fuzzy decisions according to generalized modus ponens (GMP) and generalized modus tollens (GMT) [58,66,67,103,112,118]. Unfortunately, they are often not meaningful and not efficient in real applications. We have found two main reasons which are (1) the conventional fuzzy sets theory always takes whole fuzzy sets as fundamental elements for fuzzy reasoning without intelligently splitting any fuzzy set originally defined by Zadeh [111] into more meaningful and more useful fuzzy subsets, and (2) the commonly used fuzzy reasoning methods are too logical to physically map input fuzzy sets to reasonable output fuzzy sets. For the second issue, various training algorithms such as gradient-descent algorithms, genetic algorithms and neural-nets-based algorithms have been used in fuzzy system modeling [42,43,72,78,103,115]. However, these algorithms still have disadvantages which are (1) for both given data and a required error, they cannot effectively find the appropriate number of fuzzy rules for the corresponding error, i.e., they cannot guarantee that they can successfully train a fuzzy system with the given number of fuzzy rules to satisfy the required error and (2) it is difficult for these algorithms to find optimal or near-optimal solutions because they may reach local minima.

For the above important problems in fuzzy reasoning and fuzzy system

11.2 A Normal-Fuzzy-Reasoning-Based Fuzzy System

Since Zadeh first defined the concept of fuzzy set in 1965 [111], fuzzy sets representing linguistic concepts such as *very high* and *very low* have been used as basic computational elements for making fuzzy reasoning and processing fuzzy information. The data granularity of conventional fuzzy rule bases with fuzzy sets is much lower than that of databases with raw data. However, "how to select optimal data granularity to generate knowledge bases ?" is still unsolved.

Since the data granularity of conventionally used fuzzy sets is too low to contain more heuristic information and potential knowledge, we use new primary fuzzy sets with higher data granularity than commonly used fuzzy sets to construct primary fuzzy rule bases which can contain more heuristic knowledge and useful information.

A crisp-input-crisp-output fuzzy system is described below,

IF x_1 is A_1^k and ... and x_n is A_n^k THEN y is B^k

x_1 is a_1^* and ... and x_n is a_n^*

Conclusion : $\qquad\qquad\qquad\qquad\qquad\qquad y$ is b

where a_i^* for $i = 1, 2, ..., n$ are crisp values, and b is a crisp value.

For clarity, the normal fuzzy reasoning method is proposed as follows.
Step 1: Calculate the strengths of firing rules,

$$\lambda^k = [\prod_{i=1}^{n} \mu_{A_i^k}(a_i^*)]^{1-\gamma^k + \frac{\gamma^k}{n}}, \qquad (11.1)$$

where γ^k are the compensatory degrees.
Step 2: For simplicity, only two typical cases are discussed below, the others can similarly be dealt with.
CASE 1: If x_i increases, then y will increase.
If $a_i^* \leq a_i^k$

$$Then \quad \mu_{A_i^k}^l(a_i^*) = \mu_{B^k}^l(b_i^{*k}), \qquad (11.2)$$

$$Else \quad \mu_{A_i^k}^r(a_i^*) = \mu_{B^k}^r(b_i^{*k}). \qquad (11.3)$$

CASE 2: If x_i increases, then y will decrease.
If $a_i^* \leq a_i^k$

$$\text{Then } \mu_{A_i^k}^l(a_i^*) = \mu_{B^k}^r(b_i^{*k}), \tag{11.4}$$

$$\text{Else } \mu_{A_i^k}^r(a_i^*) = \mu_{B^k}^l(b_i^{*k}). \tag{11.5}$$

Step 3: Calculate the average fuzzy cardinalities

$$\theta^k = \frac{1}{n}\sum_{i=1}^n \mu_{A_i^k}(a_i^*). \tag{11.6}$$

Step 4: Calculate the average expected values of output crisp values. The two useful methods are given below,
(1) the weighted method:

$$\delta^k = \frac{\sum_{i=1}^n b_i^{*k}\mu_{A_i^k}(a_i^*)}{\sum_{i=1}^n \mu_{A_i^k}(a_i^*)}, \tag{11.7}$$

(2) the average method:

$$\delta^k = \frac{1}{n}\sum_{i=1}^n b_i^{*k}. \tag{11.8}$$

Step 5: The normal defuzzification method (NDM) is used to calculate the output expected value b. The two useful NDMs are given below,
(1) The fuzzy-cardinality-weighted method:

$$b = \frac{\sum_{i=1}^n \delta^k \theta^k \lambda^k}{\sum_{i=1}^n \theta^k \lambda^k}, \tag{11.9}$$

(2) The firing-strength-weighted method:

$$b = \frac{\sum_{k=1}^m \delta^k \lambda^k}{\sum_{k=1}^m \lambda^k}. \tag{11.10}$$

11.3 Various Single-Input-Single-Output (SISO) fuzzy systems

Here we discuss an SISO fuzzy system with m fuzzy IF-THEN rules such that

$$IF\ x\ is\ A_k\ THEN\ y\ is\ B_k, \tag{11.11}$$

Sec. 11.3. Various Single-Input-Single-Output (SISO) fuzzy systems

where x and y are input and output fuzzy linguistic variables respectively, fuzzy linguistic values A_k and B_k are defined as follows,

$$\mu_{A_1}(x) = \begin{cases} \dfrac{a_2 - x}{a_2 - a_1} & \text{for } a_1 \le x \le a_2 \\ 0 & \text{otherwise,} \end{cases} \quad (11.12)$$

$$\mu_{A_j}(x) = \begin{cases} \dfrac{x - a_{j-1}}{a_j - a_{j-1}} & \text{for } a_{j-1} \le x \le a_j \\ \dfrac{a_{j+1} - x}{a_{j+1} - a_j} & \text{for } a_j \le x \le a_{j+1} \\ 0 & \text{otherwise,} \end{cases} \quad (11.13)$$

$$\mu_{A_m}(x) = \begin{cases} \dfrac{x - a_{m-1}}{a_m - a_{m-1}} & \text{for } a_{m-1} \le x \le a_m \\ 0 & \text{otherwise,} \end{cases} \quad (11.14)$$

$$\mu_{B_k}(y) = \begin{cases} 1 + \dfrac{y - b_k}{\eta_k} & \text{for } (b_k - \eta_k) \le y \le b_k \\ 1 - \dfrac{y - b_k}{\eta_k} & \text{for } b_k \le y \le (b_k + \eta_k) \\ 0 & \text{otherwise,} \end{cases} \quad (11.15)$$

where $j = 2, 3, ..., m-1$, a_k and b_k are centers of these triangular membership functions of x and y, respectively, η_k are widths of membership functions of y for $k = 1, 2, ..., m$.

Takagi-Sugeno's SISO fuzzy system is described below,

$$\text{IF } x \text{ is } A_k \text{ THEN } y^k = p_0^k + p_1^k x, \quad (11.16)$$

where p_i^k for $i = 0, 1$ and $k = 1, 2, ..., m$ are parameters.

Since the parameters p_i^k for $i = 0, 1$ and $k = 1, 2, ..., m$ in Takagi-Sugeno's fuzzy system have no physical meanings related to the output membership functions, a conventional Takagi-Sugeno's fuzzy system can not be constructed directly from the given high-level fuzzy rules.

According to [103], Wang's SISO fuzzy system $\bar{g}(x)$ for the triangular membership functions is a linear system such that

$$\bar{g}(x) = b_k \frac{a_{k+1} - x}{a_{k+1} - a_k} + b_{k+1} \frac{x - a_k}{a_{k+1} - a_k}, \quad (11.17)$$

where $x \in [a_k, a_{k+1}]$ for $k = 1, 2, ..., m-1$.

The centers of the output membership functions are used in the Wang's SISO fuzzy system, but the widths of the output membership functions are not used in it. So the Wang's SISO fuzzy system cannot completely describe the whole fuzzy system.

According to the normal fuzzy reasoning method in section 11.2.2, we can construct a normal fuzzy system $g(x)$ directly based on given fuzzy rules. For clarity, we show the procedure step by step:

Step 1: Choose $\gamma^k = 0$. Because only two fuzzy rules are fired each time, we have strengths of firing rules:

$$\lambda^k = \frac{a_{k+1} - x}{a_{k+1} - a_k}, \tag{11.18}$$

$$\lambda^{k+1} = \frac{x - a_k}{a_{k+1} - a_k}, \tag{11.19}$$

for $k = 1, 2, ..., m-1$.

Step 2: Now $x \in [a_k, a_{k+1}]$ for $k = 1, 2, ..., m-1$, therefore we have

(1) If the output y is an increasing function, then

$$\lambda^k = 1 - \frac{y_k - b_k}{\eta_k}, \tag{11.20}$$

$$\lambda^{k+1} = 1 + \frac{y_{k+1} - b_{k+1}}{\eta_{k+1}}. \tag{11.21}$$

We have the two corresponding outputs of x:

$$y_k = b_k + \eta_k \lambda^{k+1}, \tag{11.22}$$
$$y_{k+1} = b_{k+1} - \eta_{k+1} \lambda^k. \tag{11.23}$$

(2) If the output y is a decreasing function, then

$$\lambda^k = 1 + \frac{y_k - b_k}{\eta_k}, \tag{11.24}$$

$$\lambda^{k+1} = 1 - \frac{y_{k+1} - b_{k+1}}{\eta_{k+1}}. \tag{11.25}$$

Sec. 11.4. Universal approximation 157

We have the two corresponding outputs of x:

$$y_k = b_k - \eta_k \lambda^{k+1}, \qquad (11.26)$$

$$y_{k+1} = b_{k+1} + \eta_{k+1} \lambda^k. \qquad (11.27)$$

Finally, we have

$$y_k = b_k + w\eta_k \lambda^{k+1}, \qquad (11.28)$$

$$y_{k+1} = b_{k+1} - w\eta_{k+1} \lambda^k, \qquad (11.29)$$

where $w = 1$ and $w = -1$ are for an increasing output function and a decreasing output function, respectively.

Step 3 to 5: According to the rest steps in the normal fuzzy reasoning method in Section 11.2, we have the normal fuzzy system $g(x)$ such that

$$g(x) = \frac{y_k \lambda^k + y_{k+1} \lambda^{k+1}}{\lambda^k + \lambda^{k+1}}. \qquad (11.30)$$

Finally, we have

$$g(x) = b_k \frac{a_{k+1} - x}{a_{k+1} - a_k} + b_{k+1} \frac{x - a_k}{a_{k+1} - a_k} + w(\eta_k - \eta_{k+1}) \frac{(a_{k+1} - x)(x - a_k)}{(a_{k+1} - a_k)^2}, \qquad (11.31)$$

where $x \in [a_k, a_{k+1}]$ for $k = 1, 2, ..., m-1$.

Obviously, Wang's fuzzy system $\bar{g}(x)$ is just the linear part of our new system $g(x)$.

11.4 Universal approximation

In this section, we will prove that the normal fuzzy logic system $g(x)$ given by Eq. (11.34) can approximate any continuous function $f(x)$ in a compact interval $[a, b]$.

We first recall that any continuous function $f(x)$ in a compact interval $[a, b]$ is uniformly continuous [85]. Therefore, for arbitrary $\varepsilon > 0$ there exist $\alpha_i \in [a, b]$ with

$$a = \alpha_1 < \alpha_2 < ... < \alpha_n = b, \qquad (11.32)$$

such that

$$\mid f(x) - f(\alpha_i) \mid < \varepsilon, \qquad (11.33)$$

$$|f(x) - f(\alpha_{i+1})| < \varepsilon, \qquad (11.34)$$

whenever $x \in [\alpha_i, \alpha_{i+1}]$ for $i = 1, 2, ..., n-1$.

Lemma 11.1: The Wang's SISO fuzzy system $\bar{g}(x_i)$ in the form of Eq. (11.20) can approximate $f(x)$ in an arbitrary small degree.

Proof: For arbitrary $\varepsilon > 0$, there exist $a_i \in [a, b]$ with

$$a = a_1 < a_2 < ... < a_n = b, \qquad (11.35)$$

such that

$$|f(x) - f(a_i)| < \frac{\varepsilon}{4}, \qquad (11.36)$$

$$|f(x) - f(a_{i+1})| < \frac{\varepsilon}{4}, \qquad (11.37)$$

whenever $x \in [a_i, a_{i+1}]$ for $i = 1, 2, ..., n-1$. It follows that

$$\begin{aligned}
|f(x) - \bar{g}(x)| &= |f(x) - f(a_i)\frac{a_{i+1} - x}{a_{i+1} - a_i} - f(a_{i+1})\frac{x - a_i}{a_{i+1} - a_i}| \\
&\leq |f(x) - f(a_i)|\frac{a_{i+1} - x}{a_{i+1} - a_i} + |f(x) - f(a_{i+1})|\frac{x - a_i}{a_{i+1} - a_i} \\
&< \frac{\varepsilon}{4}.
\end{aligned} \qquad (11.38)$$

That is, $\bar{g}(x)$ can approximate $f(x)$ uniformly.

Q.E.D.

Theorem 11.1: The normal fuzzy logic system $g(x)$ in the form of Eq. (11.34) can approximate an arbitrary continuous function $f(x)$ in the closed interval $[a, b]$.

Proof: Suppose $M = max_{x \in [a_i, a_{i+1}]} |\bar{g}(x) - g(x)|$. For any $x \in [a_i, a_{i+1}]$, we have

$$\begin{aligned}
M &= |w| |\eta_i - \eta_{i+1}| max_{x \in [a_i, a_{i+1}]} \frac{(a_{i+1} - x)(x - a_i)}{(a_{i+1} - a_i)^2} \\
&= |\eta_i - \eta_{i+1}| \frac{1}{4}.
\end{aligned} \qquad (11.39)$$

When

$$|\eta_i - \eta_{i+1}| \leq 3\varepsilon, \qquad (11.40)$$

Sec. 11.5. A piecewise nonlinear constructive algorithm

we have that
$$max_{x \in [a_i, a_{i+1}]} | \bar{g}(x) - g(x) | \leq \frac{3\varepsilon}{4}. \tag{11.41}$$

According to Lemma 1, it follows that
$$| f(x) - g(x) | \leq | f(x) - \bar{g}(x) | + | \bar{g}(x) - g(x) | < \varepsilon. \tag{11.42}$$

That is, $g(x)$ can approximate $f(x)$ uniformly.

Q.E.D.

11.5 A piecewise nonlinear constructive algorithm

Suppose that there are N data pairs (x_i, y_i) for $i = 1, 2, ..., N$, $b_1 = y_1$ and $\eta_1 = |y_2 - y_1|$. The optimization function is given by:

$$Q = \sum_{i=1}^{N} [y_i - g(x_i)]^2, \tag{11.43}$$

where

$$g(x_i) = b_1 \frac{x_N - x_i}{x_N - x_1} + b_2 \frac{x_i - x_1}{x_N - x_1} + w(\eta_1 - \eta_2) \frac{(x_N - x_i)(x_i - x_1)}{(x_N - x_1)^2}. \tag{11.44}$$

We have

$$\frac{\partial Q}{\partial b_2} = 2 \sum_{i=1}^{N} [y_i - g(x_i)] \frac{x_i - x_1}{x_N - x_1} = 0, \tag{11.45}$$

$$\frac{\partial Q}{\partial \eta_2} = -2 \sum_{i=1}^{N} [y_i - g(x_i)] w \frac{(x_N - x_i)(x_i - x_1)}{(x_N - x_1)^2} = 0. \tag{11.46}$$

After solving the equations (11.48) and (11.49), we have

$$b_2 = \frac{\sum_{i=1}^{N} F_i(y_i - H_i) \sum_{i=1}^{N} G_i^2 - \sum_{i=1}^{N} G_i(y_i - H_i) \sum_{i=1}^{N} F_i G_i}{\sum_{i=1}^{N} F_i^2 \sum_{i=1}^{N} G_i^2 - (\sum_{i=1}^{N} F_i G_i)^2}, \tag{11.47}$$

$$\eta_2 = \frac{\sum_{i=1}^{N} G_i(y_i - H_i) \sum_{i=1}^{N} F_i^2 - \sum_{i=1}^{N} F_i(y_i - H_i) \sum_{i=1}^{N} F_i G_i}{\sum_{i=1}^{N} F_i^2 \sum_{i=1}^{N} G_i^2 - (\sum_{i=1}^{N} F_i G_i)^2}, \tag{11.48}$$

where

$$F_i = 1 - \mu, \qquad (11.49)$$
$$G_i = w\mu(\mu - 1), \qquad (11.50)$$
$$H_i = b_1\mu - G_i\eta_1, \qquad (11.51)$$

where

$$\mu = \frac{x_N - x_i}{x_N - x_1}, \qquad (11.52)$$

for $i = 1, 2, ..., N$.

The piecewise nonlinear constructive algorithm is described below:

(Assuming that there are N data pairs $(x[i], y[i])$ for $i = 1, 2, ..., N$ and the required maximum square error is E_m.

k: the counter of fuzzy rules.

p: the pointer to point the first data pair $(x[p], y[p])$ for the current non-linear regression function.

q: the counter of used data pairs.

n: the counter of current data pairs used in the nonlinear regression.

E_c: the current square error.

E: the total square error.

$\bar{E}$: the mean square error.

w: the current parameter for the normal fuzzy system.

η_0: the width of the current triangular output membership function.

η: the width of the coming triangular output membership function.

β_0: the center of the current triangular output membership function.

β: the center of the coming triangular output membership function.)

Step 1: Initialization: $k = 0$, $p = 1$, $\beta_0 = y[1]$, $\eta_0 = |y[2] - y[1]|$, $w = 1.0$, $E = 0.0$, $b[0] = \beta_0$ and $c[0] = \eta_0$, $d[0] = x[1]$.

Step 2: Set: $n = 1$, $k = k + 1$.

Step 3: $n = n + 1$.

If $(n = 2)$ Then set: $\beta = y[p + 1]$ and $\eta = \eta_0$.

Sec. 11.5. A piecewise nonlinear constructive algorithm

Else calculate:

$$\beta = \frac{\sum_{i=p}^{p+n} F_i(y[i] - H_i) \sum_{i=p}^{p+n} G_i^2 - \sum_{i=p}^{p+n} G_i(y[i] - H_i) \sum_{i=p}^{p+n} F_i G_i}{\sum_{i=p}^{p+n} F_i^2 \sum_{i=p}^{p+n} G_i^2 - (\sum_{i=p}^{p+n} F_i G_i)^2}, \quad (11.53)$$

$$\eta = \frac{\sum_{i=p}^{p+n} G_i(y[i] - H_i) \sum_{i=p}^{p+n} F_i^2 - \sum_{i=p}^{p+n} F_i(y[i] - H_i) \sum_{i=p}^{p+n} F_i G_i}{\sum_{i=p}^{p+n} F_i^2 \sum_{i=p}^{p+n} G_i^2 - (\sum_{i=p}^{p+n} F_i G_i)^2}. \quad (11.54)$$

If $(\eta \leq 0)$
Then set $w = -w$, and calculate:

$$\beta = \frac{\sum_{i=p}^{p+n} F_i(y[i] - H_i) \sum_{i=p}^{p+n} G_i^2 - \sum_{i=p}^{p+n} G_i(y[i] - H_i) \sum_{i=p}^{p+n} F_i G_i}{\sum_{i=p}^{p+n} F_i^2 \sum_{i=p}^{p+n} G_i^2 - (\sum_{i=p}^{p+n} F_i G_i)^2}, \quad (11.55)$$

$$\eta = \frac{\sum_{i=p}^{p+n} G_i(y[i] - H_i) \sum_{i=p}^{p+n} F_i^2 - \sum_{i=p}^{p+n} F_i(y[i] - H_i) \sum_{i=p}^{p+n} F_i G_i}{\sum_{i=p}^{p+n} F_i^2 \sum_{i=p}^{p+n} G_i^2 - (\sum_{i=p}^{p+n} F_i G_i)^2}. \quad (11.56)$$

Where

$$F_i = 1 - \mu, \quad (11.57)$$
$$G_i = w\mu(\mu - 1), \quad (11.58)$$
$$H_i = \beta_0 \mu - G_i \eta_0, \quad (11.59)$$

where

$$\mu = \frac{x[p+n] - x[i]}{x[p+n] - x[p]}, \quad (11.60)$$

for $i = p, p+1, ..., p+n$.

Step 4:

Then calculate n values of the current nonlinear regression function:

$$f[i] = \beta_0 * \mu + \beta * [1 - \mu] + w * (\eta_0 - \eta) * \mu * [1 - \mu], \quad (11.61)$$

where

$$\mu = \frac{x[p+n] - x[i]}{x[p+n] - x[p]}, \quad (11.62)$$

where $i = p, p+1, ..., p+n$. Finally, calculate the current square error:

$$E_c = \sum_{i=p+1}^{p+n} (f[i] - y[i])^2. \quad (11.63)$$

Step 5: Set: $q = p + n$. Calculate the total square error:

$$E = E + E_c. \tag{11.64}$$

Then calculate the mean squared error:

$$\bar{E} = \frac{E}{q}. \tag{11.65}$$

If $(\bar{E} \leq E_m)$ and $(\eta > 0)$

Then store parameters of the current nonlinear regression function: $a[k-1] = w$, $b[k] = \beta$, $c[k] = x[p+n]$ and $d[k] = \eta$.

Step 6: If $(q \leq N)$ and $(\bar{E} \leq E_m)$ and $(\eta > 0)$

Then goto *Step 3*.

Else goto *Step 7*.

Step 7: Set: $p = q - 2$, $\beta_0 = \beta$ and $\eta_0 = \eta$.

Step 8:

If $(q \leq N)$

Then goto *Step 2*.

Else goto *Step 9*.

Step 9: The final constructed fuzzy system is given by:

$$\begin{aligned}g(x) =\;& b[i] * \mu(x) + b[i+1] * [1 - \mu(x)] + \\ & a[i] * (d[i] - d[i+1]) * \mu(x) * [1 - \mu(x)],\end{aligned} \tag{11.66}$$

where

$$\mu(x) = \frac{c[i+1] - x}{c[i+1] - c[i]}, \tag{11.67}$$

where $x \in [c[i], c[i+1]]$ for $i = 0, 1, ..., k-1$.

The constructed input membership functions are defined by:

$$\mu_{A_0}(x) = \begin{cases} \dfrac{c[1] - x}{c[1] - c[0]} & \text{for } c[0] \leq x \leq c[1] \\ 0 & \text{otherwise,} \end{cases} \tag{11.68}$$

$$\mu_{A_j}(x) = \begin{cases} \dfrac{x - c[j-1]}{c[j] - c[j-1]} & \text{for } c[j-1] \leq x \leq c[j] \\ \dfrac{c[j+1] - x}{c[j+1] - c[j]} & \text{for } c[j] \leq x \leq c[j+1] \\ 0 & \text{otherwise,} \end{cases} \tag{11.69}$$

$$\mu_{A_k}(x) = \begin{cases} \dfrac{x - c[k]}{c[k+1] - c[k]} & \text{for } c[k] \leq x \leq c[k+1] \\ 0 & \text{otherwise,} \end{cases} \quad (11.70)$$

where $j = 1, 2, ..., k - 1$.

The constructed output membership functions are defined by:

$$\mu_{B_j}(y) = \begin{cases} 1 + \dfrac{y - b[j]}{d[j]} & \text{for } (b[j] - d[j]) \leq y \leq b[j] \\ 1 - \dfrac{y - b[j]}{d[j]} & \text{for } b[j] \leq y \leq (b[j] + d[j]) \\ 0 & \text{otherwise,} \end{cases} \quad (11.71)$$

where $j = 0, 1, ..., k$.

Step 10: End.

Since $\bar{g}(x_i)$ is a special case of $g(x_i)$ when $w = 0$, so we can use the above nonlinear constructive algorithm to model Wang's SISO fuzzy system $\bar{g}(x_i)$ by setting $w = 0$. The constructive algorithm for modeling Wang's SISO fuzzy system is called the linear constructive algorithm for convenience. For simplicity, we don't describe the linear constructive algorithm because it is just a simplified version of the new nonlinear constructive algorithm.

11.6 Simulations

In this section, three typical examples are used to verify the new nonlinear constructive approach. For convenience, we use $\bar{N}$ and $\tilde{N}$ to denote the number of fuzzy rules generated by the linear approach and the number of fuzzy rules generated by our nonlinear approach, respectively.

11.6.1 A nonlinear function approximation

We consider a simple function $y = (x-5)^2$ on $[0, 10]$. Suppose we have 101 data pairs (x_k, y_k) for $x_k = k/10$, $y_k = (x_k - 5)^2$ and $k = 0, 1, 2, ..., 100$. Our new nonlinear constructive approach needs only 1 fuzzy rule for the different square errors shown in Table 11.1. Simulation results shown in Figures 11.1 and 11.2 and Table 11.1 can indicate that our new nonlinear constructive approach is better than the linear constructive approach.

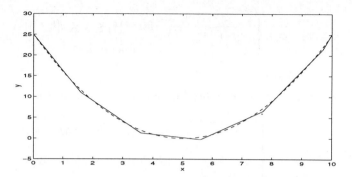

Figure 11.1. Desired values (dashed line) generated by our normal fuzzy system using 1 fuzzy rule and predicted values (solid line) generated by Wang's fuzzy system using 6 fuzzy rules when the square error is 0.1.

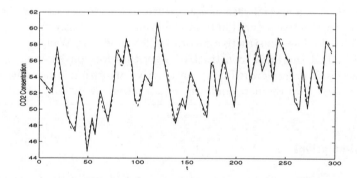

Figure 11.2. Desired values (dashed line) and predicted values (solid line) generated by the linear constructive approach with 48 fuzzy rules when the square error is 0.25.

Table 11.1: Comparison between our new nonlinear constructive approach and the linear constructive approach for the example in Section 11.6.1.

Mean Square Error	$\bar{N}$	$\tilde{N}$
0.1	6	1
0.01	10	1
0.005	11	1
0.001	17	1

11.6.2 Box and Jenkins's gas furnace model identification

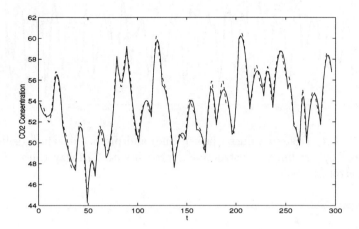

Figure 11.3. Desired values (dashed line) and predicted values (solid line) generated by the new nonlinear constructive approach with 27 fuzzy rules when the square error is 0.25.

Box and Jenkins's gas furnace data [11] is a frequently used benchmark for checking performances of fuzzy logic identification algorithms. The data set has 296 pairs of inputs (the gas flow rates) and outputs (the concentrations of CO_2). The simulation results shown in Table 11.2 and Figure 11.3 can support that our fuzzy system is able to perform complex nonlinear tasks by using piecewise nonlinear functions. Therefore, our new nonlinear constructive approach is an efficient way to make a fuzzy system with the particular triangular membership functions.

Table 11.2: Comparison between our new nonlinear constructive approach and the linear constructive approach for the example in Section 11.6.2.

Mean Square Error	$\bar{N}$	$\tilde{N}$
0.25	48	27
0.1	58	34
0.05	74	40
0.01	121	56
0.005	145	68
0.001	203	94

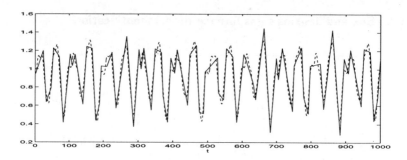

Figure 11.4. Desired values (dashed line) and predicted values (solid line) generated by the linear constructive approach with 72 fuzzy rules when the square error is 0.005.

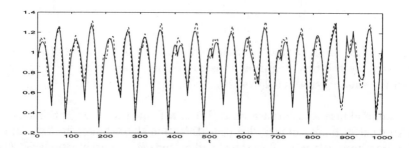

Figure 11.5. Desired values (dashed line) and predicted values (solid line) generated by the new nonlinear constructive approach with 28 fuzzy rules when the square error is 0.005.

11.6.3 A chaotic system identification

The chaotic Mackey-Glass (MG) differential delay equation is defined below:

$$\dot{x}(t) = \frac{0.2x(t-\tau)}{1+x^{10}(t-\tau)} - 0.1x(t). \tag{11.72}$$

The chaotic time series ($0 \leq t \leq 2000$) is generated by the MG equation (11.75) when $x(0) = 1.2$, $\tau = 17$ and $x(t) = 0$ for $t < 0$. For comparison, we used 1000 data pairs (t_i, x_i) for $t_i = i$ and $i = 0, 1, ..., 999$ in simulations. The simulation results shown in Table 11.3 and Figures 11.4-11.7 can strongly demonstrate that the normal fuzzy system is able to perform chaotic tasks by

Sec. 11.6. Simulations

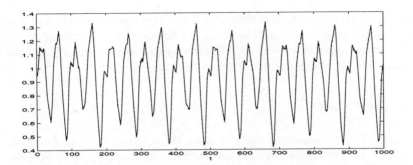

Figure 11.6. Desired values (dashed line) and predicted values (solid line) generated by the linear constructive approach with 157 fuzzy rules when the square error is 0.0001.

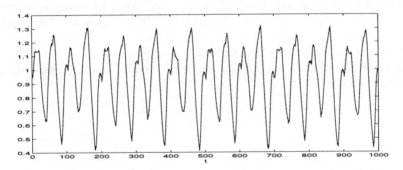

Figure 11.7. Desired values (dashed line) and predicted values (solid line) generated by the new nonlinear constructive approach with 81 fuzzy rules when the square error is 0.0001.

Table 11.3: Comparison between our new nonlinear constructive approach and the linear constructive approach for the example in Section 11.6.3.

Mean Square Error	$\bar{N}$	$\tilde{N}$
0.05	33	14
0.01	57	26
0.005	72	28
0.001	94	52
0.0005	102	62
0.0001	157	81

using piecewise nonlinear functions and it can also approximately find critical convex or non-convex curves. Therefore, the new piecewise nonlinear constructive approach provides us with an efficient way to make a normal fuzzy system with the specific triangular membership functions.

11.7 Conclusions

The normal fuzzy reasoning method is an efficient way to map fuzzy inputs to a fuzzy output physically and reasonably. By overcoming weaknesses of Wang's SISO linear fuzzy system, our new piecewise nonlinear constructive algorithm can effectively model a normal fuzzy system with the near-optimal number of fuzzy rules for both given data and a required error. In addition, our new approach is capable of generating a commonly used fuzzy system with both input and output membership functions, but Wang's fuzzy system doesn't have complete output membership functions and Takagi-Sugeno's fuzzy system has only crisp outputs. Therefore, the normal-fuzzy-reasoning-based nonlinear constructive approach provides us with a powerful tool to model a normal fuzzy system piece by piece (i.e. interval by interval) for both given data and a required error.

Chapter 12

Conclusions

In this book, we have developed effective hybrid intelligent systems merging fuzzy logic, neural networks, genetic algorithms and compensatory operations, and successfully applied it to several typical application fields. At first, we proposed a new normal fuzzy reasoning methodology based on primary fuzzy sets which overcome weaknesses of conventional fuzzy reasoning methods. Secondly, we designed several fuzzy neural networks for different applications. Finally, sufficient simulations in various areas strongly supported that the novel hybrid methodology incorporating merits of relevant techniques is useful and effective.

12.1 Main Conclusions

The major conclusions are as follows:

(1) The analysis has indicated that it is not reasonable and not efficient to always take non-primary fuzzy sets as fundamental elements for fuzzy reasoning because non-primary-fuzzy-sets-based fuzzy reasoning methods may result in losing useful knowledge and heuristic information contained in a fuzzy rule base. By splitting non-primary fuzzy sets into more useful and more heuristic primary fuzzy sets, the normal fuzzy reasoning methodology provides us with an efficient way to solve fuzzy reasoning problems such as the VICO problem and the unstable problem of a cart-pole balancing system.

(2) The normal fuzzy reasoning method uses an adaptive compensatory fuzzy operator rather than conventionally used non-adaptive ones such as Min-Max operators in order to make fuzzy reasoning more adaptive and more flexible in complex applications.

(3) It is interesting to note that the Takagi-Sugeno's fuzzy system can be

constructed directly from commonly used fuzzy rules by using the new normal fuzzy reasoning method, but the conventional Takagi-Sugeno fuzzy system has to be created from given data by using a complex algorithm. In this sense, the normal fuzzy reasoning methodology establishes a relationship between Takagi-Sugeno's fuzzy system and other fuzzy systems. In addition, a normal fuzzy system can contain not only a linear combination of inputs but also a nonlinear one by applying appropriate input and output membership functions. Since the parameters of Takagi-Sugeno's fuzzy system have no physical meaning, it is very hard to heuristically initialize them to make a gradient descent algorithm find better solutions. All parameters of a FNNKD have physical meanings, so the HGLA is very efficient for training a FNNKD by heuristically initializing parameters.

(4) Jang's ANFIS can extract Takagi-Sugeno's fuzzy rules from sample data. Since Takagi-Sugeno's rules have a fuzzy IF part and a crisp THEN part, an ANFIS cannot easily extract commonly used fuzzy rules with both a fuzzy IF part and a fuzzy THEN part. Our new FNNKD is capable of learning commonly used fuzzy rules with both a fuzzy IF part and a fuzzy THEN part from given data.

(5) As a generalized framework of Wang's fuzzy system, a FNNKD is more robust and more effective than Wang's fuzzy system according to the simulations.

(6) The fuzzy knowledge rediscovery algorithm called KRA is able to re-extract heuristic fuzzy knowledge from a fuzzy rule base, and therefore makes reasonable fuzzy reasoning according to the simulations.

(7) Our new normal fuzzy system can effectively control a cart-pole balancing system based on various fuzzy rule bases, whereas Wang's fuzzy system cannot keep the cart-pole vertical because of the VICO problem.

(8) Since weights in a conventional crisp neural network and a fuzzy-operation-oriented neural network have no explicit physical meanings, they are not capable of extracting high-level knowledge from data. In order to overcome the weaknesses of conventional crisp neural networks and fuzzy-operation-oriented neural networks, we have developed a general fuzzy-reasoning-oriented fuzzy neural network called a CFNN which is able to extract high-level knowledge such as fuzzy IF-THEN rules from either crisp data or fuzzy data. A CFNN can effectively compress a big fuzzy rule base with N fuzzy rules to a corresponding small rule base with M rules ($1 \leq M < N$), and can expand an invalid sparse fuzzy rule base to a valid one.

(9) The simulation results for the fuzzy games have indicated that TFM

with normal fuzzy reasoning can make better and more reasonable moves than TOM with precise reasoning since different global strategies are taken into account by TFM. The novel fuzzy reasoning methodology is more reasonable and more useful for making fuzzy moves than the conventional one.

(10) A CFNN made by 4 dedicated FNNKDs with the HGLA is an efficient and robust neuro-fuzzy system with the abilities of discovering fuzzy knowledge from numerical data and applying trained fuzzy rules to predict complex nonlinear behaviors such as the chaotic time series and the gas furnace model identification. Therefore, a CFNN is a powerful soft computing system with both *linguistic-words*-level fuzzy reasoning and *numerical-data*-level information processing.

(11) The new genetic fuzzy nerual network is effective for recognizing different distorted patterns with good recognition rates by incorporating genetic algorithms to the fuzzy neural network.

12.2 Future Research and Development

From a theoretical point of view, there is an advantage to investigate the convergence of the hybrid learning algorithm and the universal approximation of various fuzzy neural networks. Furthermore, one may be able to design a constructive learning algorithm to directly *calculate* all parameters of a normal fuzzy system rather than heuristically *search* for them.

From an application point of view, one should investigate the application of our new fuzzy neural networks to other fields such as computing with words, intelligent agents, expert systems, knowledge discovery in databases, data mining, etc.

Bibliography

[1] H. Adeli and S.-L. Hung, *Machine Learning - Neural Networks, Genetic Algorithms and Fuzzy Systems*, John Wiley & Sons, 1995.

[2] G.-S. Andreas, *Fuzzy Rule-Based Expert Systems and Genetic Machine Learning*, Physica-Verlag, 1994.

[3] K. Asakawa and H. Takagi, *Neural networks in Japan*, Communications of the ACM, **Vol. 37**, No. 3, pp. 106-112, March 1994.

R. Axelrod, *The Evolution of Cooperation*, Penguin Books, pp. 3-54, 1984.

[4] R. Babuska, et al., *Simplification of fuzzy rule bases*, Proc. of EUFIT'96, pp. 1115-1119, Sept. 1996.

[5] A. Bastian, *Handling the nonlinearity of a fuzzy logic controller at the transition between rules*, Fuzzy Sets and Systems, **Vol. 71**, pp. 369-387, 1995.

[6] A. Bastian and M. Mukaidono, *Identification of Fuzzy Systems with Incomplete Rule Bases*, Proc. of IPMU'96, **Vol. II**, pp. 539-544, 1996.

[7] H.R. Berenji, *Neural networks and fuzzy logic in intelligent control*, Proc. of the 5th IEEE Int. Symp. on Intelligent Control, **Vol. 2**, pp. 916-920, 1990.

[8] J.C. Bezdek and S.K. Pal, *Fuzzy Models for Pattern Recognition - Methods That Search for Structures in Data*, IEEE Press, 1992.

[9] P.P. Bonissone, P.S. Khedkar and Y. Chen, *Genetic Algorithms for Automated Tuning of Fuzzy Controllers: A Transportation Application*, Proc. of the 5th IEEE Int. Conf. on Fuzzy Systems, **Vol. 1**, pp. 674-680, 1996.

[10] J. Bowen and G. Dozier, *Solving Randomly Generated Fuzzy Constraint Networks Using Evolutionary/ Systematic Hill-Climbing,* Proc. of the 5th IEEE Int. Conf. on Fuzzy Systems, **Vol. 1**, pp. 226-231, 1996.

[11] G.E.P. Box and G.M. Jenkins, *Time Series Analysis, Forecasting and Control,* San Francisco, Holden Day, 1976.

[12] S. J. Brams, *Theory of Moves,* American Scientist, **Vol. 81**, Nov. 1993.

[13] S. J. Brams, *Theory of Moves,* Cambridge University Press, 1994.

[14] S. J. Brams and Walter Mattli, *Theory of moves: overview and examples,* Conflict Management and Peace Science, **Vol. 12**, No. 2, pp. 1-39, 1993.

[15] M.Brown and C.Harris, *Neurofuzzy Adaptive Modelling and Control,* Prentice Hall, 1994.

[16] J.J. Buckley and Y. Hayashi, *Fuzzy neural networks: a survey,* Fuzzy Sets and Systems, **Vol. 66**, pp. 1-13, 1994.

[17] J.J. Buckley, P. Krishnamraju, K. Reilly and Y. Hayashi, *Genetic Learning Algorithms for Fuzzy Neural Nets,* Proc. of the Third IEEE Int'l Conf. on Fuzzy System, **Vol. 3**, pp. 1969-1974, 1994.

[18] G. A. Carpenter and S. Grossberg, *The ART of adaptive pattern recognition by a self-organizing neural network,* IEEE Computer Mag., **Vol. 21**, No. 3, pp. 77-88, 1988.

[19] G. A. Carpenter, S. Grossberg, et al., *Fuzzy ARTMAP: A Neural Network Architecture for Incremental Supervised Learning of Analog Multi dimensional Maps,* IEEE Trans. on Neural Networks, **Vol. 3**, No. 5, pp. 698-713, 1992.

[20] C.H. Chen, *Fuzzy Logic and Neural Network Handbook,* McGraw-Hill, Inc., 1996.

[21] C.-L Chen and W.-C. Chen, *Fuzzy controller design by using neural network techniques,* IEEE Trans. on Fuzzy Systems, **Vol. 2**, No. 3, pp. 235-244, Aug. 1994.

[22] W. Chiang and J. Lee, *Fuzzy Logic for the Applications to Complex Systems,* World Scientific, 1996.

[23] O. Cordon and F. Herrera, *Generating and Selecting Fuzzy Control Rules Using Evolution Strategies and Genetic Algorithms,* Proc. of the 6th International Conf. on IPMU, **Vol. II**, pp. 733-738, July 1996.

[24] D. Driankov, H. Hellendoorn and M. Reinfrank, *An Introduction to Fuzzy Control*, Springer-Verlag, 1993.

[25] L.J. Eshelman, *Proc. of the 6th Int. Conf. on Genetic Algorithms*, Morgan Kaufmann Publishers, 1995.

[26] M. Figueiredo, F. Gomide, A. Rocha and R. Yager, *Comparison of Yager's Level Set Method for Fuzzy Logic Control with Mamdani's and Larsen's Methods*, IEEE Trans. on Fuzzy Systems, **Vol. 1**, No. 2, pp. 156-159, 1993.

[27] D.B. Fogel, *Evolutionary Computation*, IEEE Press, 1995.

[28] T. Fukuda, et al., *Structure optimization of fuzzy neural network by genetic algorithm*, Proc. of the Fifth IFSA World Congress, pp. 964-967, 1993.

[29] K. Fukushima, S. Miyake and T. Ito, *Neocognitron: A neural network model for a mechanism of visual pattern recognition*, IEEE Trans. on SMC, **Vol. 13**, No. 5, pp. 826-834, 1983.

[30] K. Fukushima, *Handwritten alphanumeric character recognition by neocognitron*, IEEE Trans. on Neural Networks, **Vol. 2**, No. 3, pp. 355-365, 1991.

[31] T. Furuhashi, *Advances in Fuzzy Logic, Neural Networks and Genetic Algorithms*, Springer, 1995.

[32] S. Ghoshray, *An Efficient Grey Prediction Model By Combining Fuzzy Logic and Genetic Algorithm*, Proc. of the 6th International Conf. on IPMU, **Vol. II**, pp. 727-731, July 1996.

[33] D.E. Goldberg, *Genetic Algorithms in Search, Optimization, and Machine Learning*, Addison-Wesley Publishing Company, Inc., 1989.

[34] M.M. Gupta and D.H. Rao, *On the principles of fuzzy neural networks*, Fuzzy Sets and Systems, **Vol. 61**, pp. 1-18, 1994.

[35] I. Hayashi, et al., *Construction of fuzzy inference rules by neural network driven fuzzy reasoning and neural network driven fuzzy reasoning with learning functions*, Int. J. Approximate Reasoning, **Vol. 6**, pp. 241-266, 1992.

[36] F. Herrera and M. Lozano, *Adaptive Genetic Algorithms Based on Fuzzy Techniques*, Proc. of the 6th International Conf. on IPMU, **Vol. II**, pp. 775-780, July 1996.

[37] J.H. Holland, *Genetic algorithms*, Scientific American, pp. 66-72, July 1992.

[38] J.J. Hopfield and D.W. Tank, *Neural computation of decisions in optimization problems*, Biological Cybernetics, **Vol. 52**, pp. 141-152, 1985.

[39] S. Horikawa, T. Furuhashi and Y. Uchikawa, *On fuzzy modeling using fuzzy neural networks with the back propagation algorithm*, IEEE Trans. on Neural Networks, **Vol. 3**, No. 5, pp. 801-806, 1992.

[40] H. Ishibuchi, R. Fujioka and H. Tanaka, *Neural Networks That Learn From Fuzzy If-Then Rules*, IEEE Trans. on Fuzzy Systems, **Vol. 1**, No. 2, pp. 85-97, 1993.

[41] H. Ishigami, et al., *Structure optimization of fuzzy neural network by genetic algorithm*, Fuzzy Sets and Systems, **Vol. 71**, pp. 257-264, 1995.

[42] J.-S.R. Jang, *Fuzzy Modeling Using Generalized Neural Networks and Kalman Filter Algorithm*, Proc. of the 9th National Conf. on Artificial Intelligence (AAAI-91), pp. 762-767, July 1991.

[43] J.-S.R. Jang, *ANFIS: Adaptive-Network-Based Fuzzy Inference System*, IEEE Trans. on SMC, **Vol. 23**, No. 3, pp. 665-685, 1993.

[44] J.-S.R. Jang and Gulley Ned, *Fuzzy Logic Toolbox For Use with MATLAB*, The MATHWORKS Inc., 1995.

[45] Y. Jin, et al., *Neural network based fuzzy identification and its application to modeling and control of complex systems*, IEEE Trans. on SMC, **Vol. 25**, No. 6, pp. 990-997, June 1995.

[46] A. Kandel and S.C. Lee, *Fuzzy switching and automata: theory and applications*, Crane Russak, New York, 1979.

[47] A. Kandel, *Fuzzy Techniques in Pattern Recognition*, John Wiley, 1982.

[48] A. Kandel and G Langholz, *Hybrid Architectures for Intelligent Systems*, CRC Press, 1992.

[49] A. Kandel, A. Martins and R. Pacheco, *Discussion: On the Very Real Distinction Between Fuzzy and Statistical Methods*, TECHNOMETRICS, **Vol. 37**, No. 3, pp. 276-281, 1995.

[50] A. Kandel, Y.-Q. Zhang and T. Miller, *Knowledge representation by conjunctive normal forms and disjunctive normal forms based on n-variable-m-dimensional fundamental clauses and phrases,* Fuzzy Sets and Systems, **Vol. 76**, pp. 73-89, 1995.

[51] A. Kandel and Y.-Q. Zhang, *Axioms-based CNFs and DNFs Constructed by n-variable-m-dimensional Fundamental Clauses and Phrases,* Proc. of NAFIPS'96, pp. 41-45, June 1996.

[52] A. Kandel, Y.-Q. Zhang and T. Miller, *Fuzzy Neural Decision System for Fuzzy Moves,* Proc. of the 3rd World Congress on Expert Systems, pp. 718-725, 1996.

[53] A. Kaufmann, *Theory of Fuzzy Subsets,* Academic Press, New York, 1975.

[54] J.M. Keller and H. Tahani, *Backpropagation neural networks for fuzzy logic,* Inform. Sci., **Vol. 62**, pp. 205-221, 1992.

[55] W. J.M. Kickert, *Fuzzy theories on decision-making,* Kluwer Boston Inc., 1978.

[56] J. Kim, Y. Moon and B.P. Zeigler, *Designing Fuzzy Net Controllers Using Genetic Algorithms,* IEEE Control System, pp. 66-72, June 1995.

[57] S. Kim and G.J. Vachtsevanos, *A polynomial fuzzy neural network for identification and control,* Proc. of 1996 Biennial Conf. of NAFIPS, pp. 5-9, June 1996.

[58] G.J. Klir and Yuan Bo Yuan, *Fuzzy Sets and Fuzzy Logic Theory and Applications,* Prentice Hall P T R, 1995.

[59] T. Kondo, *Revised GMDH Algorithm estimating degree of the complete polynomial,* Trans. Soc. Instrument and Contr. Engineers, **Vol. 22**, No. 9, pp. 928-934, 1986 (in Japanese).

[60] B. Kosko, *Neural Networks and Fuzzy Systems - a dynamical systems approach to machine intelligence,* Prentice-Hall, 1992.

[61] B. Kosko, *Fuzzy Thinking: The New Science of Fuzzy Logic,* New York: Hyperion, 1993.

[62] P.V. Krishnamraju, et al., *Genetic learning algorithms for fuzzy neural networks,* Proc. of IEEE Int. Conf. on Fuzzy Systems, **Vol. III**, pp. 1969-1974, 1994.

[63] A. Krone, *Advanced rule reduction concepts for optimising efficiency of knowledge extraction*, Proc. of EUFIT'96, pp. 919-923, Sept. 1996.

[64] K. Kupper, *Self learning fuzzy models using stochastic approximation*, Proc. of the 3rd IEEE Conf. Control Applications, **Vol. 3**, pp. 1723-1728, 1994.

[65] H.K. Kwan and Y.Cai , *A Fuzzy Neural Network and its Application to Pattern Recognition*, IEEE Trans. on Fuzzy Systems, **Vol. 2**, No. 3, pp. 185-193, 1990.

[66] C.-C. Lee, *Fuzzy Logic in Control Systems: Fuzzy Logic Controller-Part I*, IEEE Trans. on SMC, **Vol. 20**, No. 2, pp. 404-418, 1990.

[67] C.-C. Lee, *Fuzzy Logic in Control Systems: Fuzzy Logic Controller-Part II*, IEEE Trans. on SMC, **Vol. 20**, No. 2, pp. 419-435, 1990.

[68] S.C. Lee and E.T. Lee, *Fuzzy sets and neural networks*, J. Cybernet., **Vol. 4**, pp. 83-103, 1974.

[69] S.C. Lee and E.T. Lee, *Fuzzy neural networks*, Math Biosci., **Vol. 23**, pp. 151-177, 1975.

[70] J.-W. Li, *Taichi and Bagua*, Tianjin University Press, 1989.

[71] C.-T. Lin and C.S.G. Lee, *Neural-Network-Based Fuzzy Logic Control and Decision System*, IEEE Trans. on Neural Networks, **Vol. 40**, No. 12, pp. 1320-1336, Dec. 1991.

[72] C.-T. Lin, *Neural Fuzzy Control Systems with Structure and Parameter Learning*, World Scientific, 1994.

[73] C.-T. Lin and C.S.G. Lee, *Neural Fuzzy Systems - A Neural-Fuzzy Synergism to Intelligent Systems*, Prentice Hall P T R, 1996.

[74] R. P. Lippmann, *An introduction to computing with neural nets*, IEEE ASSP Mag., **Vol. 4**, No. 2, pp. 4-22, 1987.

[75] G. L. Martin and J. A. Pittman, *Recognizing hand-printed letters and digits using backpropagation learning*, Neural Computation, **Vol. 3**, No. 2, pp. 258-267, 1991.

[76] D. Nauck and R. Kruse, *A fuzzy neural network learning fuzzy control rules and membership functions by fuzzy error backpropagation*, Proc. of IEEE Internt. Conf. on Neural Networks, pp. 1022-1027, 1993.

[77] O. Nelles, M. Fischer and B. Muller, *Fuzzy Rule Extraction by a Genetic Algorithm and Constrained Nonlinear Optimization*, Proc. of the 5th IEEE Int. Conf. on Fuzzy Systems, **Vol. 1**, pp. 213-219, 1996.

[78] J. Nie and D. Linkens, *Fuzzy-Neural Control Principles, Algorithms and Applications*, Prentice Hall, 1995.

[79] F.S.M. Nobre, *Genetic-neuro-fuzzy systems: a promising fusion*, Proc. of FUZZ-IEEE/IFES'95, **Vol. I**, pp. 259-266, 1995.

[80] S. K. Pal and D. K. D. Majumder, *Fuzzy Mathematical Approach to Pattern Recognition*, Wiley Eastern Ltd., 1986.

[81] W. Pedrycz, *An identification algorithm in fuzzy relational systems*, Fuzzy Sets and Systems, **Vol. 13**, pp. 153-167, 1984.

[82] W. Pedrycz, *Fuzzy neural networks and neurocomputations*, Fuzzy Sets and Systems, **Vol. 56**, pp. 1-28, 1993.

[83] W. Poundstone, *Prisoner's Dilemma*, Doubleday, 1992.

[84] T.C. Rubinson and G.M. Geotsi, *Estimation of Subjective Oreferences Using Fuzzy Logic and Genetic Algorithms*, Proc. of the 6th International Conf. on IPMU, **Vol. II**, pp. 781-786, July 1996.

[85] W. Rudin, *Principles of Mathematical Analysis*, McGraw-Hill, Inc., New York, 1976.

[86] D.E. Rumelhart, McClelland, and the PDP Research Group, *Parallel distributed processing: explorations in the microstructure of cognition*, Vol.1:Foundations. Cambridge,MA: MIT Press, 1986.

[87] R.M. Saleem and B.E. Postlethwaite, *A comparison of neural networks and fuzzy relational systems in dynamic modelling*, Int. Conf. Control '94, **Vol. 2**, pp. 431-439, 1987.

[88] G.W. S Schwede and A. Kandel, *Fuzzy maps*, IEEE Trans. on SMC, **Vol. 9**, pp. 669-674, 1977.

[89] R.N. Sharpe, et al., *A methodology using fuzzy logic to optimize feedforward artificial neural network configuration*, IEEE Trans. on SMC, **Vol. 24**, No. 5, pp. 760-768, May 1994.

[90] K. Shimojima, et al., *Self-tuning fuzzy modeling with adaptive membership function, rules, and hierarchical structure based on genetic algorithm*, Fuzzy Sets and Systems, **Vol. 71**, pp. 295-310, 1995.

[91] M. Sugeno and K. Tanaka, *Successive identification of a fuzzy model and its application to prediction of a complex system*, Fuzzy Sets and Systems, **Vol. 42**, pp. 315-334, 1991.

[92] M. Sugeno and G.T. Kang, *Structure Identification of Fuzzy Model*, Fuzzy Sets and Systems, **Vol. 28**, pp. 15-33, 1988.

[93] M. Sugeno and T. Yasukawa, *A Fuzzy-Logic-Based Approach to Qualitative Modeling*, IEEE Trans. on Fuzzy Systems, **Vol. 1**, No. 1, pp. 7-31, 1993.

[94] H. Takagi and I. Hayashi, *NN-driven fuzzy reasoning*, Int. J. Approximate Reasoning, **Vol. 5**, No. 3, pp. 191-212, 1991.

[95] T. Takagi and M. Sugeno, *Fuzzy Identification of Systems and Its Applications to Modeling and Control*, IEEE Trans. on SMC, **Vol. SMC-15**, No. 1, pp. 116-132, 1985.

[96] R.M. Tong, *The evaluation of fuzzy models derived from experimental data*, Fuzzy Sets and Systems, **Vol. 4**, pp. 1-12, 1980.

[97] I.B. Turksen, *Interval-valued fuzzy sets based on normal forms*, Fuzzy Sets and Systems, **Vol. 20**, pp. 191-210, 1986.

[98] I.B. Turksen, *Interval-valued fuzzy sets and 'compensatory AND'*, Fuzzy Sets and Systems, **Vol. 51**, pp. 295-307, 1992.

[99] I.B. Turksen, *Fuzzy normal forms*, Fuzzy Sets and Systems, **Vol. 69**, pp. 319-346, 1995.

[100] S.G. Tzafestas and A.N. Venetsanopoulos, *Fuzzy Reasoning in Information, Decision and Control Systems*, Kluwer Academic Publishers, 1994.

[101] C.-H. Wang, et al., *Fuzzy B-spline membership function (BMF) and its applications in fuzzy-neural control*, IEEE Trans. on SMC, **Vol. 25**, No. 5, pp. 841-851, May 1994.

[102] J.-H. Wang, *The Theory of Games*, Oxford University Press, 1988.

[103] L.-X. Wang, *Adaptive Fuzzy Systems and Control design and stability analysis*, PTR Prentice Hall, 1994.

[104] L. Wang and R. Langari, *Building Sugeno-Type Models Using Fuzzy Discretization and Orthogonal Parameter Estimation Techniques*, IEEE Trans. on Fuzzy Systems, **Vol. 3**, No. 4, pp. 454-458, Nov. 1995.

[105] L. Wang and R. Langari, *Complex systems modeling via fuzzy logic*, IEEE Trans. on SMC, **Vol. 26**, No. 1, pp. 100-106, Feb. 1996.

[106] P.J. Werbos, *Beyond regression: new tools for prediction and analysis in the behavioral sciences*, Ph.D. thesis, Harvard Univ., Nov., 1974.

[107] C.W. Xu, *Fuzzy systems identification*, IEE Proceedings, **Vol. 136**, Part.D, No. 4, 1989.

[108] R.R. Yager and L.A. Zadeh, *Fuzzy Sets, Neural Networks, and Soft Computing*, Van Nostrand Reinhold, 1994.

[109] R.R. Yager and D.P. Filev, *Essentials of Fuzzy Modeling And Control*, John Wiley & Sons, 1994.

[110] J. Yen, B. Lee and J.C. Liao, *Using Fuzzy Logic and a Hybrid Genetic Algorithm for Metabolic Modeling*, Proc. of the 5th IEEE Int. Conf. on Fuzzy Systems, **Vol. 1**, pp. 220-225, 1996.

[111] L.A. Zadeh, *Fuzzy Sets*, Information and Control, **Vol. 8**, pp. 338-353, 1965.

[112] L.A. Zadeh, *Outline of a new approach to the analysis of complex systems and decision processes*, IEEE Trans. on SMC, **Vol. 3**, No. 1, pp. 28-44, 1973.

[113] L.A. Zadeh, *Fuzzy Logic = Computing with Words*, IEEE Trans. on Fuzzy Systems, **Vol. 4**, No. 2, pp. 103-111, May 1996.

[114] Y-Q. Zhang, A. Kandel and M. Friedman, *Hybrid Decision-making System for Fuzzy Moves*, Proc. FUZZ-IEEE/IFES'95, Vol. II, pp. 621-626, 1995.

[115] Y.-Q. Zhang and A. Kandel, *Genetic-Guided Compensatory Neurofuzzy Systems*, Proc. of the 6th International Conf. on IPMU, **Vol. I**, pp. 181-185, July 1996.

[116] K.C. Zikidis and A.V. Vasilakos, *ASAFES2: a novel, neuro-fuzzy architecture for fuzzy computing, based on functional reasoning*, Fuzzy Sets and Systems, **Vol. 83**, pp. 63-84, 1996.

[117] H.-J. Zimmermann and P. Zysno, *Latent connective in human decision making*, Fuzzy Sets and Systems, **Vol. 4**, pp. 37-51, 1980.

[118] H.-J. Zimmermann, *Fuzzy Sets Theory and Its Applications*, Kluwer Academic Publishers, 1996.

Index

A

Activation functions, 3
Adaptive Optimization, 4
Affirmative statement, 12, 15
Algebraic product, 31
Algebraic sum, 31
ANFIS, 4
Associativity laws, 19
Axioms-based, 11

B

Boolean logic, 12, 23
Boolean truth table, 12
Bounded product, 32
Bounded sum, 32

C

Cart-pole balancing, 81
CFNN, 68
Chaotic time series, 106
Chromosomes, 64
CICO, 4, 70
CIFO, 4, 70
Clockwise progression, 122
Commutativity laws, 19
Compensatory ANDs, 10
Compensatory degrees, 46, 110

Compensatory operation, 10, 11
Compensation principles, 9-11
Complex systems, 1
Compositional rule of inference, 55
Computational intelligence, 1, 5
Computing with words (CW), 1
Conservation laws, 26, 30, 32, 35
Constructive algorithm, 159
Constructive approach, 152
Contradiction law, 11
Counterclockwise progression, 122
Crisp features, 67
Crisp weights, 3
Crossover, 64

D

Data Granularity, 1, 2
Data mining, 171
Defuzzification, 43, 71-73
DeMorgan's laws, 19
Distorted patterns, 149, 150
Divergence, 38

E

Excluded middle law, 11
Expected values, 47
Expert systems, 5

F

Feature expressions, 67
FICO, 4, 70
FIFO, 4, 70
FNNKD, 58
Fuzzy ARTMAP, 5
Fuzzy Associate Memory, 145
Fuzzy automata, 5
Fuzzy CNF, 11
Fuzzy controllers, 49
Fuzzy DNF, 11
Fuzzy IF-THEN rules, 3, 5
Fuzzy information, 1
Fuzzy logic, 1
Fuzzy logic toolbox, 104
Fuzzy map, 11
Fuzzy numbers, 1, 77
Fuzzy reasoning, 1, 4
Fuzzy rule base, 77, 83
Fuzzy rules, 2
Fuzzy sets, 1
Fuzzy games, 115
Fuzzy information capacity, 67
Fuzzy knowledge compression, 95
Fuzzy knowledge expansion, 95
Fuzzy knowledge rediscovery, 8, 71
Fuzzy neural network, 3
Fuzzy-operation-oriented, 3
Fuzzy-reasoning-oriented, 4
Fuzzy moves, 115

G

Game theory, 5
Game transformation, 129-135
Gas furnace model, 112
Gaussian quaternion, 67
General fuzzy logic, 26
Generalized modus ponens, 6
Generalized modus tollens, 6

Genetic algorithms, 1, 4, 61
Geometric expressions, 67
GFNN, 145
Global game, 116
Global PDs, 129, 140
Gradient descent, 64, 152
Granularity, 1, 2
Granule, 1

H

Heuristic algorithms, 5
Heuristic information, 3
Heuristic parameters, 52, 78
HGLA, 61
Hybrid intelligent systems, 1
Hybrid search algorithm, 4
Hybrid learning algorithm, 4

I

Information granularity, 6
Intelligent agents, 171
Imprecision, 1
Involution law, 19

K

Karnaught map, 12
Kaufmann's fuzzy truth table, 11, 13
Knowledge discovery in databases, 171
Knowledge discovery, 6, 171
KRA, 53

L

Learning algorithm, 61
Linguistic variable, 1
Linguistic terms, 3
Linguistic words, 6

Index

Local game, 116
Local PDs, 131

M

Membership functions, 4
Min and Max, 11
Model Identification, 112
Modeling, 104
Monotonically decreasing, 45
Monotonically increasing, 45
Monotonically non-decreasing, 41
Monotonically non-increasing, 41
Mutation, 64

N

Natural selection, 4
Near-optimal, 4, 168
Negative statement, 12, 15
Neural Networks, 1, 3
Neuro-fuzzy systems, 5
Non-adaptive, 169
Nonlinear function approximation, 77
Nonlinear regression, 161
Nonlinear system, 8
Non-singleton Fuzzifiers, 49
Normal fuzzy controllers, 49
Normal fuzzy reasoning, 44
Numerical data, 6

O

Optimistic, 10, 46
Optimization, 4

P

Partition, 61
Pattern recognition, 5, 145
Pessimistic, 10, 46

Precise decision systems, 123
Precise rules, 2
Prediction, 104
Primary fuzzy sets, 41, 42
Prisoners' Dilemma, 119, 127
Probabilistic reasoning, 1

S

Self-organizing learning algorithm, 147
Soft computing, 1, 5
Singleton Fuzzifiers, 49
SISO, 154
Sparse fuzzy rule base, 98

T

Taichi, 9
Takagi-Sugeno fuzzy system, 6, 58
T-conorm, 15, 35, 39
Theory of Fuzzy Moves (TFM), 115, 136
Theory of Moves (TOM), 115
Time series prediction, 106
T-norm, 15, 35, 39
Trapezoidal fuzzy set, 42
Trapezoidal-type fuzzy set, 45
Trapezoidal quaternion, 67

U

Uncertainty, 1
Universal approximation, 157
Universe of discourse, 41, 42
Universal fuzzy truth table, 15
Unnecessary nonlinearity, 72

V

Variable-Input-Constant-Output, 42

W

Washing machines, 5
Weighted, 48

Y

Yin-Yang, 9, 10
Yager's LSM, 72